CARNIVORES OF EUROPE

Robert Burton

CARNIVORES
OF
EUROPE

B. T. Batsford   London

Acknowledgements

Research into the lives of carnivores is a continuing and expanding subject. I would like to thank the following for giving their expert advice: John A. Burton, Paul Chanin, Christopher Cheeseman, Ian Linn, Gwyn Lloyd, Clifford Owen and Jane Thornback.

The author and publishers would like to thank the following for permission to reproduce black and white photographs: Ardea Photographics (Nos 2, 3, 4, 7 and 12); Jane Burton (Nos 1, 5, 6, 10 and 11); Bruce Coleman Ltd (No 8); Eric and David Hosking (No 9).

For the colour illustrations, Ardea (Stoat, Polecat and Feline genet); Bruce Coleman Ltd (Polar bear, Brown bear, Red fox, Weasel, Beech marten, Pine marten, Badger, Otter, Lynx, Raccoon dog).

First published 1979

© Robert Burton, 1979

ISBN 0 7134 0690 9

Printed in Great Britain by
Cox & Wyman Ltd.,
London, Fakenham and Reading
for the publishers
B. T. Batsford Ltd, 4 Fitzhardinge Street,
London W1H 0AH

Contents

List of illustrations

List of maps

ICELAND

SWED

Kiöl

HEBRIDES

Bergen

NORWAY

Kintyre

REP.OF
IRELAND

I.OF MAN

UNITED
KINGDOM

DENMARK

Norfolk

NETHER
LANDS

POLAND

BELGIUM

EIFEL

GERMANY

R.
Mosel

FRANCE

CZECHOSLOVAKIA

Ca...

Cantabrian
Mountains

SWITZER-
LAND

AUSTRIA

HUNGARY

PORTUGAL

Pyrenees

Massif
Central

Dolomites

RO

SPAIN

ITALY

YUGOSLAVIA

Sierra Morena

Apennines

ADRIATIC SEA

Coto Doñana

BALEARIC Is.

CORSICA

Dalmatia

BU

Majorca

ALBANIA

SARDINIA

GREECE

CRETE

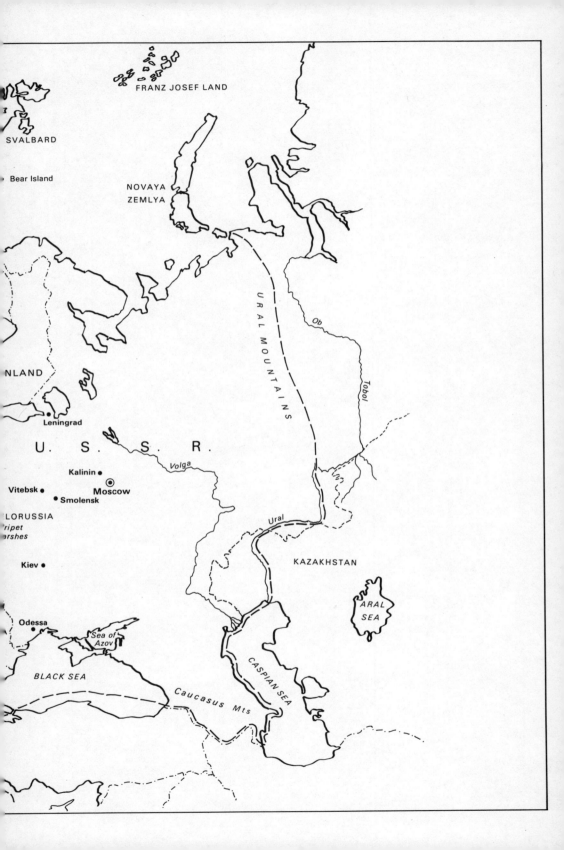

Introduction

The carnivores are hunting, flesh-eating mammals, and they are fur-bearers, so, in one way or another, they have played an important part in the lives of men. They have been hunted because of their depredations on domestic animals and because of their threat to man himself. They have been hunted for their fur and they have been chased for sport.

ANATOMY AND PHYSIOLOGY

The order Carnivora comprises a group of primarily terrestrial mammals adapted for hunting mainly warm-blooded prey. There are exceptions, like the insect-eating aardwolf of Africa and the mainly vegetarian giant panda, and it also comes as a surprise to find out how much vegetable food many flesh-eaters do consume. Nevertheless, the simple definition helps to distinguish the Carnivora from the seals of the order Pinnipedia, which were at one time classed as Carnivora.

All families of the Carnivora, except the Protelidae (aardwolf) are represented in the European fauna, although the Viverridae (Egyptian mongoose and feline genet) and the Procyonidae (raccoon) are represented only by introduced species and the Hyaenidae (striped hyaena) has only the barest foothold in the Caucasus. Overall, Europe is well supplied with carnivores in having one-half of the species that occur in the Palaearctic region. There are ten times as many species living throughout the world but, as may be expected, the majority live in tropical regions where the fauna is more diverse.

The families of carnivores are grouped into two superfamilies based on characters of the skull and male reproductive organs. They need little introduction as they are easily recognizable by their external appearance without the need for examination of the teeth and skeletal characters, which form the main basis of mammalian systematics.

ORDER CARNIVORA

Superfamily Canoidea
Family Canidae
Family Ursidae
Family Procyonidae
Family Mustelidae

Superfamily Feloidea
Family Viverridae
Family Hyaenidae
Family Protelidae
Family Felidae

11

In anatomy and physiology, the carnivores are designed for the capture and killing of prey. Most are running animals but many climb, dig or swim as well, although none are so adapted for these traits as a monkey, a mole or a dolphin. The form of the skeleton is, in fact, generalized with only relatively minor alterations to suit a particular way of life. These alterations are centred in the limbs and the skull, the two parts of the skeleton which show the greatest adaptive variation in mammals and are, hence, used by systematists for classification of species. They are also the parts of the carnivore which are used for capture and killing of prey so it is instructive, on these two counts, to give them a brief examination. The skull has two functions. It is literally the 'nerve centre' of the animal through containing the brain and the main sense organs. The brain is large as hunting demands intelligence and flexibility of behaviour. Secondly, the skull is the site for mastication, the initial preparation of food, and the overall shape of the skull is very much determined by the nature of the masticatory process, which involves the teeth and the jaw mechanism.

Teeth in different parts of the mouth have different functions which is reflected in their form. At the front of the mouth lie the chisel-shaped incisors used for cutting. Behind them lie the long, pointed canines or fangs and along the length of the jaw there are the cheek-teeth—the premolars and molars—which are used for chewing. The principal features of carnivore dentition are the size of the canines and the development of the cheekteeth. The canines are developed into sharp lances for piercing prey, to hold and tear its flesh (compare them with the blunt 'eye-teeth' in humans). They are particularly well developed in the cats which kill their prey by forcing the canine teeth between the bones of the neck. The form of cheekteeth depends on the carnivore's diet. Both premolars and molars bear points or cusps which are arranged so that upper and lower teeth intermesh to slice or chew food. They are, however, relatively small because meat needs little predigestion with salivary enzymes. The carnivore only needs to cut its food small enough to pass down its gullet. Many carnivores have a reduction in the number of cheekteeth, as shown by the dental formulae in the Table. There may be a further reduction through the loss of small premolars in life. Two teeth on each side of the mouth are further specialized for shearing flesh. The last premolar of the upper jaw and the first molar of the lower jaw bear blade-like cusps which slide past each other in a scissoring action. Called carnassials, these specialized teeth are best developed in the cats, weasel and stoat. The omnivorous bears and badgers have reduced cusps on all their cheekteeth and the flattened molars are used for giving fibrous vegetable matter the extra crushing it needs before digestion.

	incisor	canine	premolar	molar
Basic mammal	$\frac{3}{3}$	$\frac{1}{1}$	$\frac{4}{4}$	$\frac{3}{3}$
Canidae	$\frac{3}{3}$	$\frac{1}{1}$	$\frac{4}{4}$	$\frac{2}{3}$
Raccoon dog	$\frac{3}{3}$	$\frac{1}{1}$	$\frac{4}{4}$	$\frac{2\text{–}3}{4}$
Ursidae*	$\frac{3}{3}$	$\frac{1}{1}$	$\frac{4}{4}$	$\frac{2}{3}$
Procyonidae	$\frac{3}{3}$	$\frac{1}{1}$	$\frac{4}{4}$	$\frac{2}{2}$
Mustelidae	$\frac{3}{3}$	$\frac{1}{1}$	$\frac{4}{4}$	$\frac{1}{2}$
Otter	$\frac{3}{3}$	$\frac{1}{1}$	$\frac{3\text{–}4}{3}$	$\frac{1}{2}$
Viverridae	$\frac{3}{3}$	$\frac{1}{1}$	$\frac{4}{4}$	$\frac{2}{2}$
Felidae	$\frac{3}{3}$	$\frac{1}{1}$	$\frac{3}{2}$	$\frac{1}{1}$

Dental formulae of carnivores. The two rows of figures in each formula represent the numbers of teeth in upper and lower jaws.

* The first premolar is sometimes shed.

The impact of a carnivore striking with its opened jaws and the forces engendered by tearing and chewing flesh necessitate strong bones and powerful muscles. The two main jaw muscles are the temporalis, running from the side of the braincase to the top of the coronoid process of the lower jaw (the vertical 'fin' behind the teeth), and the masseter, running from the zygomatic arch (the 'handle' between the eye socket and the back of the skull) to the bottom of the coronoid process. Their action serves rather different ends. The anterior fibres of the temporalis, which run almost horizontally to the back of the skull, work most effectively when the jaws are wide open, whereas the posterior fibres of the temporalis and the masseter exert their maximum force when the jaws are almost shut. The anterior part of the temporalis, therefore, is most useful for driving the canines into the prey's flesh, while the masseter and the posterior part of the temporalis power the chewing premolars and molars, particularly the slicing carnassials.

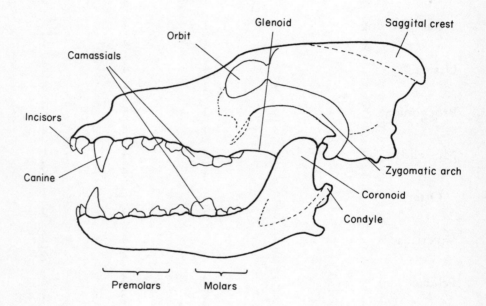

An idea of the strength of the jaw muscles is given by the size of the zygomatic arch and the area of attachment of the temporalis muscles on the braincase. In some carnivores, notably the badger and the bears, this area is increased by the sagittal crest, a ridge of bone on the top of the skull. Contraction of these muscles, along with other, lesser muscles, also causes the jaws to twist and slip from side to side in their sockets. These movements allow the animal to adjust the positioning of upper and lower teeth for efficient chewing, but, to prevent the jaws being pulled too far, the ball of the jaw, the condyle, is firmly held in a deep socket, the glenoid.

In mustelids, the skull is long and low with a consequent highly developed posterior part of the temporalis which exerts a massive dislocating force on the jaw articulation. This is counteracted by the development of bony flanges holding the jaw in place. When a stoat's jaws are nearly shut, and the temporalis muscles are exerting their maximum force, they will not drop away from the glenoid. This condition is carried further in badgers and wolverines where the flanges prevent the jaws from being dislocated in any position. Ewer (1973) points out that this solid hinge does not mean that these animals have an especially powerful bite; it is merely a means of coping with the unusual arrangement of the jaw muscles, and the tenacity of a badger's bite is in some measure due to the fact that the muscles are working best when the jaws are nearly closed.

The limbs of carnivores fall into two categories. The hind feet of bears and all the feet of the raccoon family are plantigrade. The whole of the hind foot rests on the ground, as in man. The rest are digitigrade; they stand 'on tiptoe' with the heel off the ground. (Bears have

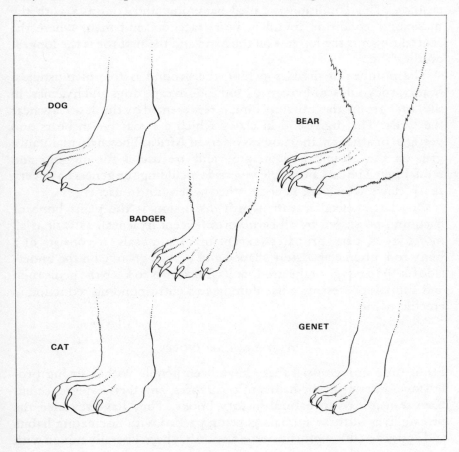

digitigrade front feet.) The broad feet and mobile ankles and wrists of the plantigrades carry them easily over broken country and, despite their heavy build, they are sure-footed. The digitigrade mammals are designed for speed. Standing on tiptoe increases the effective length of the limb and hence the length of its reach so allowing greater speeds to be attained. The joints become more restricted so that they hinge to any extent only in a fore–aft axis and each limb has become a lever for throwing the animal forward in bounds. Some flexibility has to be retained in the interests of securing prey, so the limbs have not become so rigid as those of hoofed mammals where the lower limb bones are fused together. With increasing speed, the feet suffer a pounding at each stride and the running carnivores, such as the dogs and cats, take the weight of the body principally on the two central toes which are the longest and are bound together by ligaments to make a solid cushion. This condition is called paraxonic and contrasts with the mesaxonic condition found in bear, raccoon, and man, where the central finger is the longest on the hand and the first toe is the longest on the foot.

The number of digits touching the ground is five in mustelids, bears, procyonids and viverrids and four in cats, dogs and hyaenas. In the latter group, the 'missing' digit is represented by the dew claw near the wrist. The digits end in claws which are massive in bears and vestigial or absent in the clawless otters of Africa. They help maintain a grip on the ground and on prey and are useful for digging and scratching. The cats and some viverrids including the genets have very sharp claws which are kept sheathed when not in use.

One last skeletal feature worth discussion is the penis bone or baculum, possessed by all carnivores (except hyaenas), as well as all insectivores, bats, primates (except man) and seals. It consists of a bony rod, often elaborately shaped and bears a groove on the underside for the passage of the urethra. Its function is to keep the penis rigid and animals possessing a baculum have a corresponding reduction in erectile tissue.

STUDYING CARNIVORES

From time immemorial there have been people with a strong professional interest in the habits of carnivores, and their experience has been transmitted to natural history books. They have revealed the private lives of these animals as being packed with fascinating habits and aptitudes, but, unfortunately, fiction has been liberally mixed with the facts. Knowledge has all too often consisted of anecdotes in which the narration did not distinguish between observation and deduction. There was, for instance, a belief that weasels sucked the blood of their prey, a supposition based on the finding of bloody puncture wounds on the necks of their victims. We now know that a bite to the neck is the typical *coup de grâce* delivered by the weasel and its relatives. It is aimed at the most vulnerable part of the body and is an instinctive pattern of behaviour.

This is typical of the recent developments in the study of carnivores. The old naturalists were excellent observers but they were sometimes rather uncritical in the conclusions they drew from their findings. They were liable to plump for the most appealing solution without putting it to the test. The modern zoologist subjects his observations to the most rigorous investigation so that the discursive anecdote is now replaced by a scientific paper detailing the minutiae of experiments and objective observation. Perhaps something has been lost by this approach. The image of the patient observer being rewarded by the exciting glimpse into an animal's private life has been replaced by dry scientific experiment and analysis; the countryman in plus-fours has given way to the researcher in white lab-coat. Yet this is by no means always the case in the study of carnivores. An increasing trend in the last few years has been to study carnivores in the wild and many carnivores live where the country still is very wild. Despite the use of infra-red television cameras to watch wild animals undisturbed at night and the strapping of radio transmitters to the backs of animals to follow their movements, the hours of patient waiting or chance observation of the traditional naturalist are still very much needed to study these elusive and shy animals.

The approaches of the systematic scientist and the patient naturalist are complementary, so that distinction between the two becomes blurred. J. D. Lockie (1966) has studied the social life of weasels by conferring individuality through punched patterns of holes in their ears. Over a period of one year, the weasels were monitored by catching them in a series of permanently positioned traps. The range of each weasel was thereby indicated by the traps it habitually visited. Not all the weasels were permanent inhabitants of the area. Some were caught only once and were dubbed 'transients' because they were merely passing through, and the conclusion was drawn that some weasels hold territories which they defend against intruders, such as the homeless transients. The bare bones of this notion was given substance by the chance observation of a weasel escorting another from its territory in a pell–mell, shrieking chase. The pursuit ended at a known territory boundary and the pursuing weasel was caught and identified as the owner of that territory. Armed with the results of this systematic approach we can go back to the anecdotes of earlier natural-ists and interpret their observations. Mortimer Batten (1920) describes how he and his brother followed one weasel as it chased another for a full five minutes and we can conjecture that Batten was seeing a territorial dispute because the chase flowed up and down a road where, presumably, the boundary between the two territories ran.

Despite advances in scientific techniques, it is often necessary to rely on anecdotal natural history for information on the private lives of carnivores. The breeding of a species can be studied by dissecting the reproductive organs from a series of corpses to give precise infor-mation on time of mating, duration of pregnancy, fertility and so on, but the dissections give no idea of how the sexes meet, of how the male woos the female and of how she makes her choice from the available suitors. For those whose interest lies in studying the behaviour of animals, the selection of a mate is one of the most important aspects of an animal's reproductive life, but information is virtually restricted to the anecdotes of the patient naturalist or casual observer, and to observations on captive animals. Studies of captive animals may be undertaken with the rigour of scientific study but it must always be borne in mind that captivity restricts the full expression of an animal's behaviour. Nevertheless, we shall see that a great deal has been learned from captive animals.

To summarize, the study of carnivores, and of other animals, pro-ceeds in three ways. There is the rigorous scientific examination of wild animals, by trapping, tracking or dissection, for instance; a simi-larly rigorous observation of captive animals; and the slow accumu-lation of anecdotal information collected by lucky chance but carefully scrutinized before its acceptance as hard fact. From the zoologist's point of view the anecdote is the least satisfactory, because it is hard to repeat, yet it stimulates scientific enquiry by showing facets of the animal's life that need to be investigated and it adds an excitement to zoology by revealing tantalizing glimpses of mysterious habits. Vesey-Fitzgerald (1942) has given an account of a badger funeral. The female badger dug a grave and, with the help of another male, dragged the body of her mate into it and covered it with earth. What specu-lation this conjures! Are all badgers, or at least mated badgers, buried like this? Is it merely an act of hygiene on a par with the well-known use of latrines, and how did the female communicate her need for assistance? This sort of behaviour cannot be subjected to experimental study, and light can be shed on the matter only by further chance observations.

The systematic long-term study by professional zoologists has challenged some of the ideas which have become embedded in carnivore lore. It used to be boldly stated that otters moved upstream with migrating salmon, but no evidence was offered and none is forthcoming from intensive studies. It is quite likely, however, that local otters are attracted to salmon spawning grounds in due season but it is unlikely, from what is now known of the otters' habits, that they would follow the fish. Similarly, it is held that badgers bring their bedding out of the sett to give it an airing, but I have not found a detailed description of the piles of bedding lying outside the entrance being taken in again. The old bedding is turfed out and replaced with fresh material.

Apart from the study of carnivores for their own interest, research has been prompted because many species are of economic importance. Here, either the aim is to promote carnivores so they can be cropped for their fur, or ways are sought to eliminate them because they are pests. The latter is the most frequent reason for research as control needs to be based on a sound knowledge of the pest's biology if it is to to be carried out efficiently. The impact of the pest must be assessed, and its numbers and ability to maintain those numbers must be evaluated. This requires research into three main aspects of the pest's life: its diet, its spatial organization (how the population is dispersed over the countryside) and its breeding.

Because carnivores impinge on human affairs by eating animals which man considers his own, a great deal of research into carnivores has been aimed at studying diet and feeding habits. This area of study is comparatively easy to undertake because sampling is simple. For example, otters are rarely seen, but their droppings are seen on conspicuous places. They can be collected and analysed to show the nature of the otter's diet. Foxes are hardly less elusive but they are trapped or shot by keepers and farmers. The contents of their stomachs can then be preserved and examined. In this way a picture is gradually built up of an animal's diet, and it is often possible to show whether a carnivore is merely hunting whatever is most readily caught or whether it has a real preference for certain prey. If sufficient samples are collected, trends can be discerned: the beech marten's preferences for rodents and small birds in the summer and the female pine marten's failure to catch such large prey as the male. The practical consequences of these findings are to show what threat a carnivore really does pose to human interests, as opposed to the supposed threat of folklore and conjecture.

The most commonly used method of estimating the importance of particular items in the diet is to record the frequency that they occur in droppings or stomachs, expressed as a percentage. While this method will show how one item, say voles or insects, varies in the diet through the year or from place to place, in itself it is misleading for comparison of the different items. The occurrence of one insect receives the same score as the occurrence of one or even several voles, which will have contributed far more to the carnivore's nutrition. To overcome this problem, estimates of the mass of each prey can be made by multiplying the number of times an animal occurs in a sample by its live weight.

Eventually it is possible to estimate very roughly how many voles, beetles, sheep or whatever are eaten by a carnivore per meal or per year. When this has been accomplished, the next step is to find out how many carnivores are hunting these animals so that the total impact of the predator can be estimated. This involves making a census of the population, and there is no easy way to do this. Carnivores are secretive and well-scattered over the countryside so it is difficult to count heads. Moreover, numbers are continually changing through breeding, mortality and migration. An intimate knowledge of the population is, therefore, required and this is where the study of carnivores becomes particularly interesting from the point of view of the general naturalist interested in the private life of the animal. Marking with eartags or other devices turns an anonymous group of animals into personalities whose actions can be followed over a period of time.

A general understanding of the social and spatial organization of carnivores is emerging from the many studies involving individually identified animals. The area over which the animal habitually moves becomes established by repeatedly capturing an individual in a network of permanently positioned live-traps or through following its movements by its tracks or an attached radio transmitter, and when a series of animals is studied, in this way their relationships, particularly between individuals of the same or opposite sex, become clearer.

Each established member of the population has an area where it lives—its home ground where it sleeps, hunts, eats, drinks and may mate and raise a family. This area ranges from a few hectares, or less, to several square kilometres depending on the body size of the species, the abundance of food and, sometimes, competition for the available space. If the animal makes migrations, the area will be vast. The area where the animal carries out its routine everyday activities is called its home range. It may expand in winter when the animal has to forage farther abroad for food, and contract in summer when food is abundant.

Within the range there may be core areas which are used more regularly. The animal may occupy one core area for a few days, then move to the next; then criss-crossing the range will be tracks, often so well used as to be obvious to the human eye. Sometimes the home range is not so much an area of ground enclosed within a boundary but, similar to a railway network, a series of paths connecting important places such as dens, feeding and drinking places. H. G. Lloyd (1975) has compared a fox's range with a string of beads: the beads are feeding areas, the string a pathway leading through them. With this sort of range two or more animals can, to human eyes, exist in the same place yet be leading separate, solitary lives, except for sharing a trail at different times.

At one time the home range was simply referred to as the animal's territory, but this term has come to mean a defended area, such as the territory of a bird, an area of ground defended by singing and displaying during the nesting season. There are some mammals which defend a territory against their own kind in this manner but others are tolerant and several individuals share an area, although there may be a social hierarchy of rank among them. In certain species it appears that part of the home range, perhaps the core areas, are defended and occupied exclusively by one animal, while peripheral areas are shared with neighbours (Jewell, 1966). Whatever the form of its range the occupant makes forays beyond its normal limits to determine the disposition of its neighbours, to discover whether females are available for mating or whether the disappearance of a neighbour will allow an expansion of the home range. Altogether, there is an impression of fluidity, that an animal's movements are subject to change as circumstances alter, rather than tied within the confines of a rigid boundary.

Communication between neighbours and the defence of their property is not so obvious in mammals as it is in the songs and displays of birds. Face-to-face encounters seem to be rare and communication is effected by voice and scent. The importance of the latter in mammalian communication has been appreciated only relatively recently. The carnivores leave note of their presence and passage by means of scent marks in the form of faeces, urine and the secretions of special glands. As an animal goes about its daily round, it deposits its scent at intervals and loses no opportunity to investigate the scent marks of others. Analysis of these scents is only now being attempted but it is probable that an individual's scent denotes its identity, sex and perhaps reproductive status.

Outside the society of established animals, there are the wandering transients, like the weasels described earlier. Because they are caught once and never seen again, it is difficult to find exactly who the transients are. They could, perhaps, be 'trap-shy' animals which are seldom, if ever, caught despite being resident in an area where trapping is taking place. It also seems that some subordinate animals can live in the territory of a dominant individual by adopting a 'low profile', but most transients are young animals dispersing from the maternal home. If they are to survive and breed, each must find an unoccupied area where it can set up its own range. Young carnivores may disperse considerable distances. The average for foxes in Welsh sheep country is 13·4 kilometres for dog-foxes and 2·2 kilometres for vixens (Lloyd, H. G., 1976) but individuals may wander much farther under certain circumstances and turn up many kilometres beyond the species' usual range. The practical significance of dispersal of carnivores lies in the speed with which an area can be recolonized after a control programme has eliminated them. Dispersal is an especially important factor in the current spread of rabies across Europe. For example, killing foxes, the species most implicated in the transmission of rabies, may actually hasten the spread of the disease because fox society is disrupted and foxes are induced to wander farther.

Pest control is ineffective if an animal breeds faster than it is being killed off, so its breeding potential must be established. From the same animals which gave the stomach contents for analysis of diet, are taken the reproductive organs for the investigation of the reproductive processes. Reproductive physiology in the form of development of testes and ovaries shows the time of year at which mating can take place and an examination of the uterus shows the size of the litter and timing of birth. Longevity linked with the birth of young gives the productivity of the species.

In H. G. Lloyd's Welsh foxes, vixens give birth to a mean of 4·7 cubs a year but 20·5 per cent in any year fail to give birth at all. Many of these cubs will fail to contribute much to future generations because, of 100 foxes surviving their first year of life, only forty-three see their second birthday, and only one or two survive to six years. Over half the population consists of sub–adults. Life expectancy is usually short in wild animals and is further reduced by control measures, as in these Welsh foxes. Yet the species can defend itself by raising its productivity, perhaps by increased fertility, giving birth to larger litters or by better survival of cubs. In an American study, it was estimated that even if 68 per cent of foxes in New York State were killed in one year the population would have recovered completely by the next year, (Layne and McKeon, 1956). The conclusion is that if a carnivore is numerous to start with, only a very vigorous control programme has any chance of success.

Figures of pregnancy rates and adult survival are not the most exciting reading. As in a study of a carnivore's feeding habits, where the technique of catching prey is more interesting than its percentage in the diet, observations of reproductive behaviour are more exciting than statistics of reproductive physiology. Eavesdropping on the family life of carnivores is, however, very difficult and it is necessary to rely largely on observation of captive or semi-captive animals.

There is one problem where reproductive physiology cannot be properly reconciled with reproductive behaviour. In the normal process of mammalian reproduction the fertilized ovum divides repeatedly in the uterus and develops into a ball of tissue—the blastocyst. The blastocyst becomes firmly embedded, or implanted, in the lining of the uterus and a two-way passage of substances is set up across the placenta. Food and oxygen pass into the developing embryo and its waste products diffuse back into the maternal tissues. But in some mammals implantation of the blastocyst is delayed for weeks or months. This is known to happen in seals, in roe deer, in the nine-banded armadillo and, among the carnivores, in the stoat, badger, brown bear and others.

The function of delayed implantation is not altogether clear. In general, young animals are born at the most appropriate time of the year for their survival. Delayed implantation could be a means of stretching the period of pregnancy so that the birth occurs at the most auspicious time, yet allowing mating to take place when it is most convenient for the sexes to come together and there is time to spare for the rigmarole of courtship. In the case of the brown bear, the value of delayed implantation seems to be quite clear. Birth takes place in the winter den so that the cubs are old enough to accompany their mother when she emerges in spring. Gestation would last only about six weeks if it were not prolonged by delayed implantation. So, mating takes place in high summer, when life is easy, rather than in the dead of winter. There is no obvious explanation for delayed implantation in other species and it is very difficult to explain the very long delay in the stoat when the closely related weasel has no delayed implantation.

While research into the behaviour of carnivores is beginning to give a good insight into the way that they arrange their lives and their position in the ecological set-up of the countryside, a note of caution must be introduced. The results of one investigation must not be taken as being necessarily representative of the species as a whole. The diet of an animal reflects what is available in its habitat. It is unwise to state boldly that otters eat trout, for example. They feed on marine fishes when they live along the seashore and on trout if they inhabit a trout stream. Even the social organization of a species varies with circumstances. For instance, weasels in Scottish plantations are organized in a different way from those living in an Oxford wood (p. 83); and it is not clear whether foxes are naturally monogamous or polygamous. Local conditions probably dictate the social system, as they do in human society, so it is impossible to make hard and fast, sweeping statements about carnivore life.

The foregoing paragraphs cannot be considered as an exhaustive description of the study of carnivores, which would be out of place here. The intention is to highlight some of the areas in which research is proceeding and which have contributed to piecing together the life stories described in the following chapters. Paradoxically, the studies which are used to formulate control programmes can also be used to help conservation, of which there is urgent need, because effective control and destruction of habitat have combined to drive some carnivores to the wall, threatening their very existence. For example, the wolf, harried for centuries on all sides, now requires urgent protective measures if it is to survive.

Because the systematic study of carnivores is still relatively new, our present knowledge of all species is patchy and of some species extremely scant. The level of knowledge much depends on the importance of the species in human affairs, and this is reflected in the following pages. For example, the ubiquitous red fox commands many pages; the Arctic fox very few. The red fox is the more studied of the two because it is a pest to farmers all over Europe and is the main carrier of rabies in Europe (Lloyd, H. G., 1976). It is therefore vital to understand its life so that it may be attacked at its weakest point. But not much is known about the Arctic fox, because it is limited to the northern fringes of Europe where it does little harm. This is a pity, both because the social life of an Arctic fox must be as intrinsically fascinating as that of a red fox, and because it is able to survive the polar winter.

Of all the six continents, Europe is the least well defined. It might even be fair to say that, had the world's geographers not been confined largely to Europe, first in Classical times and again after the Renaissance, there would not have been a continent of Europe. The region we know as 'Europe' would be no more than a part of the great Asian landmass, with the same standing as the Indian peninsula. Such a scheme is used by zoogeographers who recognize the Palaearctic Region as a combination of Europe and Asia, with the area south of the Himalayas hived off as the Oriental Region but with North Africa and the Near East included.

Europe is more a human cultural unit which is why books are devoted to European mammals, birds, insects and plants, as well as to history, art, architecture and so on. This presents some difficulties when considering European fauna because the area of European culture cannot be easily marked out. Russia was once under Asian influence and Europe was considered to stop at the western border of Russia, but the situation is now reversed and European-orientated Russian influence spreads into Asia. The natural boundary between Europe and Asia is undoubtedly the ridge of the Ural Mountains, but this peters out in the Kirghiz Steppes, leaving an ill-defined border between the mountains and the Black Sea. In the eighteenth century, the south-eastern border of Europe was considered to run up the Sea of Azov and the River Volga. Currently it is held to run farther east, so that the Caucasus Mountains, including Georgia, Armenia and Azerbaijan, and the western coast of the Caspian Sea make up the extreme south-east corner of Europe. This still leaves a gap between the Urals, and atlases are vague about filling this gap. Political maps show Kazakhstan as Asian but there is an inclination for the geographical boundary to run up the Ural River.

At the opposite, north-west, corner of the continent, there is cause for dissension about the inclusion of the offshore islands in Europe, but the situation is clearer here. Novaya Zemlya can be seen as a continuation of the Urals, while Franz Josef Land, Svalbard and Iceland lie in limbo between Europe and America but are politically tied to Europe.

To what extent does the problem of defining Europe affect a discussion of carnivores? The northern islands are inhabited by only two carnivores: the polar bear and the Arctic fox. The former especially has only a slender foothold on the mainland and is, perhaps, only a marginal European species, while the latter penetrates deep into Scandinavia. In the south-east corner the disputed zone is more complicated. It has been usual for authors to avoid the issue by confining themselves to mammals of western Europe. Miller's *Catalogue of the Mammals of Western Europe* (1912) and Corbet's *The Terrestrial Mammals of Western Europe* (1966) are limited to Europe west of the Russian border, and van den Brink uses the 30° E. line of longitude in *A Field Guide to the Mammals of Britain and Europe* (1967). This line runs just west of Odessa, Kiev, Vitebsk and Leningrad. In following such courses, these authors do, in fact, omit only a handful of species living in the Caucasian region.

The Checklist of Palaearctic and Indian Mammals (Ellerman and Morrison-Scott, 1951) shows the following to have a European distribution confined to the Caucasus: striped hyaena, jungle cat, Pallas's cat, leopard and tiger. The last is described as being no more than a visitor from neighbouring Iran and the others are essentially Asian mammals with barely a foothold in Europe. They can be dismissed as being of little importance in a European context, as compared with the jackal which spreads more extensively into south-eastern Europe, and they are consequently omitted from this book.

Leaving aside these borderline species, the checklist of European carnivores is still complicated by unresolved taxonomic problems, namely as to whether certain forms constitute two species or subspecies of a single species. Those in dispute are lynx and polecats of the genus *Putorius* and maybe, according to some, even wildcat and weasel. Here I follow Corbet's (1966) suggestion of adopting the Russian distinction of two polecats (European *M. putorius* and steppe *M. eversmanni*) with a third species, the marbled polecat (*Vormela peregusna*), and of uniting the boreal and pardel lynxes in a single species. Other authorities support combining the lynxes into a single species, but this is contested by van den Brink (1970). The wild cats of Sardinia, Corsica and Majorca used to be classed as *Felis libyca*, the African wildcat (Ellerman and Morrison-Scott, 1951), but the African wildcat is now considered to belong to the same species as the European wildcat *F. sylvestris* (Corbet, 1966). In addition there have been persistent reports of a second, smaller, species of weasel but these have not been substantiated.

This treatment gives nineteen species of undeniably native European carnivores. To this total must be added three recently introduced species—American mink, raccoon and raccoon dog—and two species—genet and mongoose—in all likelihood anciently introduced. Human intervention can also be seen in the establishment of feral populations of domestic dogs, cats and ferrets. On the debit side, the lion and the sable have been driven out of Europe in historic times.

Several European carnivores extend their ranges well beyond the boundaries of this continent, into Africa, Asia and North America, and for some Europe is on the limits of their range and their European populations form only a small fraction of their numbers. Where possible, accent is placed on studies of the European population, on the Scandinavian wolverine rather than the more numerous North American population, for instance. This accent on Europe is to be preferred—except where European studies are lacking—because it is very likely that the biology of a species varies in detail between continents. (It is not even safe to assume that it will behave in the same way in different parts of one continent.) Moreover, a lengthy account of an animal that is scarce in Europe would be inappropriate. The American wolf must be as well known as the European red fox but it would be out of place to devote too much space to the handful of European wolves.

The Dog Family
(Canidae)

The dogs are the best-known animals. Domestic dogs are familiar as house pets and have been bred and trained for many purposes. They have also been used extensively in physiological research. Members of the dog family are usually long-legged and they are capable of running long distances before tiring. They are digitigrade with four toes on the hind feet and five on the forefeet, although one of these, the dew claw, does not touch the ground. The claws are blunt and cannot be retracted. The tail is usually long and tends to be carried parallel with or lower than the back. The head, on the other hand, is usually erect, in contrast to the cats' heads. The head also differs from the cats' in its shape: the muzzle is long and the braincase relatively large. The ears are erect.

The family Canidae can be divided into three main groupings: the genus *Canis* including wolf, coyote, the jackals and the domestic dog; the foxes of the genus *Vulpes* with the Arctic fox *Alopex* and fennec *Fennecus*; and the New World dogs of the genera *Dusicyon, Urocyon* etc. A few species do not fall into any of these categories; the bush dog of South America, the Indian dhole, the African hunting dog and the bat-eared fox.

The dogs are generalized hunters, showing little specialized anatomy or behaviour. The canine teeth are not particularly sharp and there is no well orientated death bite as in some other carnivores. Some dogs are solitary, some hunt in pairs, while some species work in family parties or packs. Prey is hunted by stealth or run down in a long chase. Although live prey makes up the bulk of the family's diet many dogs scavenge or eat vegetable food. The dog family is the most vocal of the carnivores, individuals communicating with a variety of barks, howls, snarls and whines. There is usually a single litter born per year with up to a dozen or more pups or cubs.

Wild dogs are found all over the world except in Madagascar, New Zealand, Borneo, Philippines, New Guinea and many Pacific Islands. The dingo of Australia was introduced by early man. European members of the dog family comprise the wolf, red fox, Arctic fox and golden jackal. The raccoon dog has been introduced and domestic dogs sometimes become feral.

RED FOX
(Vulpes vulpes)

Proverbially cunning, the fox is the most successful of European carnivores. It has survived, and even flourished, through years of persecution and has adapted to the spread of human settlement to the extent that it is now an urban pest.

The sharp muzzle, bright eyes and erect ears give the fox its alert appearance and, no doubt, have contributed to its reputation for sagacity and the embroidery of anecdotes about its intelligence. The fox is surprisingly small and delicately built, and a well-grown dog-fox stands no more than 35–40 centimetres at the shoulder, with a head and body length of 58–77 centimetres.* The tail adds a further 32–48 centimetres. The weight varies considerably through the year, from 6–10 kilograms—a maximum of 14·3 kilograms has been recorded for a Norwegian fox. The vixen is a few centimetres shorter than the dog-fox and the average weight is about 1 kilogram less. She also has a narrower face as she lacks the distinctive cheek ruffs of the dog-fox, but this distinction is less marked in Scandinavian foxes. The paws of foxes are small and delicate and the prints are more elongated than those of a dog of similar size.

The coat is typically russet, sometimes sandy red or ginger, with greyish-white underparts and black on the ears, and very often on the front of the legs. The tail is bushy, usually tipped with white, and there is considerable variation in colour. Both the underparts and tail are rarely black. Albinos and melanics have been recorded and the escape of American silver foxes from fox farms has resulted in this form occurring in the wild in some parts of Europe. During the moult, in April and May or later, the fox undergoes a marked change in appearance. The overall colour becomes paler and the loss of hair on the body makes the legs look longer. The tail is a pathetic relic of the magnificent brush but, by October, the coat is restored to a pristine condition of dense, richly coloured fur.

The red fox inhabits Europe and Asia, from Ireland to the Bering Straits and Japan. At one time northern North America was thought to be inhabited by a distinct but very similar species, *Vulpes fulva*, but

---

*To allow comparison between species, measurements of body size are taken from van den Brink (1967).

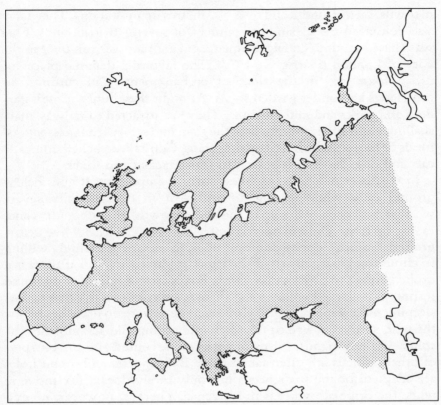

New and Old World red foxes are now regarded as the single species *Vulpes vulpes*. Foxes from Europe and North America are quite dissimilar but they represent the ends of a chain of slight variations running across the three continents (Churcher, 1959). The Eurasian fox ranges from the southern island of Novaya Zemlya in the north to Arabia, central India, northern Burma and northern Vietnam, but is absent from Iceland and Crete. It is also found in the Nile Valley and in Morocco, Algeria and Tunisia, since North Africa has a wildlife closer to Europe than to the region south of the Sahara. The European fox was introduced to Australia and North America to provide early colonists with quarry for hunting, not for the control of rabbits in Australia as is often believed. In America, the European fox was introduced to areas then south and east of the native fox's range.

Within its enormous natural and colonized range, the red fox has shown its adaptability by occupying a diversity of habitats. Foxes are found in forests, on farmland, moors and steppes, the edges of deserts, in towns, on the tundra and over 500 metres up mountains. They have also colonized the urban environment of several British cities. One was chased around Trafalgar Square and another was run over in the City of London (Harris, S., 1977). The favourite denning places of urban foxes are in the sides of embankments and cuttings, in cemeteries and under garden sheds. At night they hunt through gardens and parks and raid dustbins. They are attracted to railway marshalling yards and stations where they can find scraps from passengers' meals or the rats which feed on the scraps. Contrary to belief, domestic cats and dogs have little reason to fear attacks from foxes.

In whatever habitat a fox lives, cover is important. It goes below ground only while rearing a family, or in exceptionally severe weather, and usually lies up in the undergrowth, under boulders and, rarely, up trees. Only in open country do foxes habitually live below ground. Compared with domestic dogs, foxes are surprisingly willing to climb trees, particularly if the trunk is leaning over or if there is a broken bough to give an easy climb. Foxes are not infrequently found resting in pollarded willows and there is a record of a fox that had its sleeping place 9 metres up an elm tree. There were no branches below the nest, so the fox had to take a running jump and scramble up the trunk. On moorlands, cairns of rocks are favoured and the burrows of other animals are often taken over. Badger setts and rabbit holes are adapted for the fox's needs but only rarely does a fox move in while the original owner is in residence. There is, however, no evidence for the old story of the fox's dirty habits driving away the fastidious badger. In the eastern steppes, alpine marmot burrows are used, sometimes with marmot families continuing to live nearby. When the fox constructs its own earth, it chooses situations where the soil is well drained, deep and easily dug. There may be three or four entrances.

Where foxes are not disturbed, they are regularly active by day but they are by no means a diurnal animal, as may be deduced from the structure of the eyes, which are characteristic of a nocturnal animal. Like the cat, the fox has pupils which narrow to a vertical slit in the day and expand at night to collect as much light as possible. At the back of the eyeball there is a layer of guanine crystals, the tapetum, from which light which has not been absorbed as it passes through the retina is reflected back to give extra stimulation to the light-sensitive cells. The tapetum is responsible for the eyes of cats, foxes and other nocturnal mammals shining at night by reflecting light from a torch or car headlights. Narrowing of the pupil improves visual acuity, so the fox abroad by day can see quite well and is particularly good at detecting movement, but it probably does not see very clearly at night, and relies more on other senses. It will miss small moving prey but will be able to see and avoid obstacles.

Experiments with captive foxes (Osterholm, 1964) suggest that, in daylight, vision is the most important sense when hunting. At night, this role is usurped by hearing which alerts a fox to the presence of prey or danger. However, vision may still be needed for the final strike. Smell is used for close work—ground-nesting birds remain unmolested only a short distance from a fox trail for the fox's sense of smell does not extend more than a metre or so. To a large degree, however, the sense employed in hunting depends on circumstances. A fox's ears are sensitive to rather low-pitched sounds so that it quickly picks up and pinpoints the rustling of a vole in the grass but, surprisingly perhaps, it cannot accurately locate the vole's high-pitched squeaks. When voles are abundant and there is continuous movement through the herbage, the fox hunts by sight. The nose is also important for finding hidden prey. Foxes can find offal buried 60 centimetres underground and they can dig out nests of young rabbits and voles.

As with other carnivores, research on foxes has tended to concentrate on feeding habits. This is not only because the predatory habits of carnivores are deleterious to man's farming interests but also because examining the contents of droppings and the stomachs of dead specimens is one of the easiest ways of studying an elusive animal. In view of the fox's adaptability it is not surprising that the results of such studies differ from place to place. At the least they show that the fox has a very varied diet, but where studies can compare what foxes have been eating with what is available, it is possible to determine whether foxes select particular prey and whether predation by foxes has any effect on prey populations.

The bulk of the diet is small mammals: rabbits, hares, mice and voles; and there are clear preferences among foxes in general and by individual foxes. Nyholm (1971) records that Finnish foxes rarely eat anything but mountain hares. Captive foxes may go on hunger strike when faced with unfamiliar food and only starvation forces them to change their habits. This is unexpected in an animal which is said to be so adaptable.

Rabbits were the main food of British foxes before they disappeared under the onslaught of myxomatosis. After the rabbit population had abruptly dropped the foxes turned more to smaller mammals. Among these there is a clear preference. Field voles are preferred to bank voles and both are preferred to mice. There is the possibility that the difference in the numbers of each species eaten is a function of abundance and ease of capture. For instance, bank voles may be difficult to catch because they live in dense thickets or clumps of brambles, whereas field-voles live in open ground. Wood mice live in extensive burrow systems, where they must be fairly safe, although foxes do sometimes dig them out. Nevertheless, real preference for eating the flesh of certain prey animals has been confirmed by presenting captive foxes with various dead mammals. Lund (1962) and Macdonald (1977) have found that foxes prefer field voles to bank voles and both to wood mice if given a choice. Roger Burrows (1968), who studied a small population of wild foxes, found that brown rats were rarely eaten although they were abundant in his study area. Yet, elsewhere, rats are a preferred food. In Ireland, where there are no field voles, brown rats are the most common prey after hares and rabbits (Fairley, 1969).

Birds do not usually feature prominently in diets of foxes as shown by stomach or dropping analysis, although ground-nesting birds are vulnerable and sometimes taken. This category includes ducks, waders, gulls, pheasants, partridges and grouse. There is probably a steady toll taken of these birds where they are at risk to fox predation. However, some birds have nesting habits which minimize the risk of predation. For instance, common gulls living on the Scottish moors nest in safety on islands in lochs, and grouse, which lack such protection, are eaten only in small numbers because their nests are well spaced out and beautifully camouflaged so that foxes find them only by accident. The birds in most danger are intensively reared gamebirds and poultry. The amount of poultry and gamebirds lost to foxes in south-east England doubled after myxomatosis and preserved estates crowded with partridges or pheasants are always at risk. Nearby earths become littered with remains that bear the trademark of the fox: the primary wing feathers sheared off neatly near the roots. Such a scene will make any keeper see red but, to redress the balance, there are plenty of stories of foxes visiting hen runs and ignoring the birds. The highest losses among birds occur when hungry cubs have to be fed, which unfortunately coincides with the time of year that birds are nesting and vulnerable. During this period the cubs may receive more birds than any other food.

The amount of birds foxes can consume is remarkable. On a black-headed gull colony of 8,000 pairs, Kruuk (1964) found that four foxes living nearby accounted for 5 per cent of the adult gulls, and sandwich terns nesting among the gulls suffered proportionately heavier losses of 12 to 15 per cent. Juvenile birds suffered even more heavily than adults. In one year adult gulls made up 2·3 per cent of the foxes' diet during the gulls' breeding season, while juvenile gulls comprised 27·8 per cent and were the most important item of food after rabbits.

It is not possible to estimate how many eggs foxes eat. Eggs do not show up in analyses of stomach contents or droppings because a fox opens an egg carefully and very little of the shell is swallowed. It delicately nibbles a small hole in the shell, perhaps steadying it between the forepaws, and laps the contents. Kruuk once saw a fox cub dipping its paws into an open egg and licking them. Eggs are usually eaten where they are found but Kruuk's foxes sometimes carried eggs to a particular feeding place—an 'egg restaurant' he called it.

It is also not easy to calculate how many sheep and other large mammals foxes eat. (A fox is too small to deal with a healthy sheep over six months old, although there are records of vulnerable calves killed and roe deer are at risk when bogged in deep snow.) Although large numbers of fox droppings contain wool, giving the impression of a huge onslaught on sheep, foxes readily eat carrion and a dead sheep will last a family of foxes for some time—they do not mind if the carcase is old and putrefied. Carrion eating may even reduce the numbers of lambs killed. Gwyn Lloyd has found that there are fewer complaints of lamb killing in poor lambing seasons, when the lambs are dying of other causes and being eaten as carrion. Objectively, the number of lambs lost at any time is very small and the farmers in three Welsh counties told Lloyd that the estimated loss to foxes was only one in two hundred. This may be negligible in terms of national sheep production but the individual farmer sees every dead lamb as a financial loss, particularly in small flocks.

The full list of food eaten by foxes is very extensive but many items are usually taken only casually, like crows, earwigs and guillemots, when a fox fortuitously happens upon them. In some cases an individual fox may form the habit of hunting a particular animal—caterpillars or earthworms perhaps. Shortage of more usual foods often forces foxes to turn to items usually ignored. In Russia, cats are eaten in times of scarcity, while shrews, which are distasteful to carnivores, are eaten in the hard winters of Russia and Scandinavia. Moles are killed and usually left but Baranovoskaya and Kolosov (1935) showed from the advanced state of digestion of mole remains in fox stomachs that they may be eaten at the beginning of a hunt when the fox is hungry.

A few squirrels and occasional carnivores (e.g. weasel and badger) are caught. Macdonald (1977) records a remarkable account of a vixen feeding her own young on fox cubs caught at a neighbouring earth. Toads and snakes are rarely touched but frogs are more acceptable, as are slugs and snails. Among insect remains, those of ground beetles and dor beetles are the most commonly found in droppings. An intriguing report is of foxes catching large moths in flight (Anon., 1971). Earthworms could be taken more frequently than is suspected because their bodies are digested very rapidly. Gwyn Lloyd suggests that the afterbirths of cattle can be an important food and suggests that even droppings of unweaned lambs are eaten.

The range of vegetable food is surprising. Foxes do not eat grapes only in fable: they are well known for raiding vineyards for the fruit. They eat other fruits, including apples and pears, sloes, bilberries, haws and strawberries, and blackberrying is a favourite pastime. The fox reaches up on its hind legs to delicately pluck the berries. If that is not sufficiently strange behaviour for a carnivore, there are records of raids on potatoes and cabbages, and an apparent partiality for silage in some areas.

Foxes employ a number of tactics to capture the many kinds of animals that occur in their diet. The technique for hunting rabbits appears to be unsophisticated. The fox simply rushes headlong through a group of rabbits in an attempt to catch one off its guard. This results in the selection of young, inexperienced rabbits or those slowed down by disease or infirmity. Hit or miss hunting of this type has a high failure rate but it must be very difficult to catch a colony of rabbits unawares except by a headlong rush. Many eyes keeping watch and an efficient warning system makes stalking rabbits a difficult task, but stalking is practised on occasion. Perhaps the increased efficiency in the exploitation of the reduced rabbit population after myxomatosis (see page 38) was due to the few remaining rabbits being unable to keep sufficient lookout which allowed the fox to abandon the head-long rush and practise the more effective stalking.

By examining tracks in the sand, Kruuk deduced that gulls were caught in three different ways. Some were merely stumbled upon by accident; most were run down in the same way as rabbits; but, on dark nights in particular, foxes stalked their victims by slinking slowly with the belly almost touching the ground. The stalk ended with a jump or a few quick steps.

The fox has inspired many stories of cunning behaviour which are not fully accepted by the scientific world. One such is the stratagem of the 'fascination display' or 'charming'. The story is that a fox, seeing a party of rabbits, starts to roll on the ground to attract their attention. Then it begins to chase its own tail and run through a repertoire of antics, while the rabbits gaze spellbound, all thoughts of flight banished. The fox continues to gambol and slowly makes its way nearer the rabbits, until a sudden leap enables it to grab the nearest. There are sufficient records of this behaviour for it not to be dismissed out of hand. Some fox specialists believe it occurs; others are sceptical. Foxes are playful animals. They indulge in gambols on their own and, if rabbits or other animals are impelled by curiosity to watch the performance, it is not beyond the fox's intelligence to learn that this is a good way of obtaining a meal.

Mice and voles spend most of their lives under herbage and the fox must seek out victims by listening to faint rustlings or by watching for the slightest movement of grass stems. The rodent is caught by a characteristic 'mouse-jump', the fox rearing up on its hind legs and springing stiff-legged forward to bring its forepaws down together on the victim. By all accounts, the mouse-jump is a very successful method of attack. The head is kept low during the jump so that the snout almost makes contact at the same time as the forepaws, and the teeth are in position to deliver the *coup de grâce*.

Typically, the fox's victim is seized by the back of the neck (Tembrock, 1957) and it may be shaken or thrown into the air and recaught. The first bite causes death by injury to the head and neck but further crushing bites may be delivered to the body. Small animals are then eaten whole; larger animals are eaten piecemeal. Before the prey is eaten, it is examined carefully, and species not favoured are rejected. Shrews and other distasteful insectivores are left unless alternative food is in short supply, and the fact that shrews are killed at all suggests that a fox does not identify its victims before attacking. It shoots first and asks questions later.

Analysis of stomach contents alone will, therefore, give a false impression of the foxes' impact on prey species, but there is a further complication in the fox's habit of killing more prey than is needed. There does not seem to be any relation between a fox's hunger, or need to feed its cubs, and the amount of prey it kills. When food is readily available, an excess may be killed. In such a situation, it often caches the surplus. It may even do so when still hungry, apparently to prevent other foxes having it. A hole is dug with the forepaws, the food is thrust in and covered over with the snout. When a nest is raided and the eggs are cached, each is removed separately and buried in a different place (Kruuk, 1964). Small rodents are also cached singly. To what extent the fox returns to its caches is not known but caches certainly can be recovered. Black-headed gulls' eggs, for example, can be located three months after burying. A captive fox belonging to David Macdonald found forty-eight out of fifty of its own caches, whereas another fox was unable to locate them. It appears that a fox remembers the rough position of the cache, then pinpoints it by smell (Macdonald, 1976).

The close attention which has been paid to the feeding habits of foxes allows some assessment of the effect of a predator on the numbers of its prey, and vice versa. When myxomatosis virtually annihilated rabbit populations, the numbers of rabbits eaten by foxes obviously dropped, but the foxes were not seriously affected—partly because they could turn to other prey. Voles became the main target but more wood mice were dug from their burrows (Lever, 1959). In southern England birds suffered heavier losses, but in highland Britain insects and voles were eaten to help make good the deficit (Lever, 1959). The foxes also preyed more intensively on the remaining rabbits. In Sweden, Englund (1965) found that, while myxomatosis had reduced the rabbit population to one-twentieth of its former level, predation by foxes had been reduced only by one-half. This means that a higher proportion of the rabbit population was being killed by foxes and it would be interesting to know what extra time and effort was involved in catching those scattered survivors. A similar situation was found in Poland by Ryszkowski (1973) where foxes take a far greater proportion of the population when voles are scarce than when they are abundant.

As a general rule foxes do not reduce the population of their prey; neither do fluctuations of prey populations greatly affect the foxes, although Russian foxes decline quite dramatically when voles are scarce. Fewer cubs are born causing a low adult population twelve months after a bad vole year (Chirkova, 1953). Moreover, food shortage leads to many vixens being temporarily barren. Serious depletion of the prey population occurs only when the foxes' diet becomes limited. Foxes reaching isolated Finnish islands wreak havoc among nesting ducks and waders (Nyholm, 1971) because there is little else to eat. Kruuk found that foxes were killing more gulls each year than were being replaced by reproduction.

A predator is, as a general rule, thrifty and kills no more than it needs for itself and its family. When sated its hunting behaviour is turned off, but in certain situations the mechanism does not work and an aimless slaughter, or surplus killing, takes place. Excess prey is not cached for later use but is merely killed and abandoned. For example, the fox is notorious for its rapacity in a hen house where it is presented with an abundance of prey which cannot escape. The chickens' frantic fluttering distracts the fox and instead of concentrating on one victim it is infected by the chickens' hysteria and lashes out wildly. Surplus killing can also occur among wild, free-living prey. Kruuk found that foxes ate only about 12 per cent of the black-headed gulls that they had killed and the overwhelming majority were left to rot. Surplus killing was most noticeable on very dark nights when the gulls sat very tight on the nest and were very loath to leave when disturbed. Consequently it was a simple matter for a fox to walk from nest to nest, delivering a single bite, not always fatal. It seems that the failure of the gulls to take flight interrupted the normal sequence of the fox's hunting behaviour.

The fox's role in the predator/prey relationship is not wholly one-sided. Swedish hunters and naturalists consider that the presence of lynxes keeps foxes from any district and Haglund's (1966) survey of feeding habits of carnivores in Sweden has revealed that lynxes track down foxes or kill them in their earths. Wolves and wolverines also eat foxes. So did prehistoric man and Oscar Wilde's charge of 'the English country gentleman galloping after a fox—the unspeakable in full pursuit of the uneatable' is at least partly unfounded.

The red fox is often cited as an example of the rather small group of monogamous mammals, where a bond is formed between the sexes and the male remains with the female to help rear the family. On fur fox farms pairs have to be allowed to spend some time together if they are to mate and a male may refuse to mate with more than one female. There are instances of captive dog-foxes showing exemplary paternal behaviour in bringing food to the family, when they are normally selfish in the extreme. Monogamy is known to occur in the wild but, at other times, a dog-fox associates polygamously with a small group of vixens and may not help with the cubs.

Dog-foxes are fertile from winter to early spring, depending on geographical location. British foxes become fecund in late November, those in Lapland not until late January. Vixens have one oestrus period each year and are on heat for three days, although they are receptive for mating over a period of three weeks. As heat approaches, the dog-fox follows her assiduously, walking or trotting a few steps in the rear and stopping to sniff her urine. According to studies of captive foxes, with support from observations in the wild, the vixen starts the relationship with hostility to the dog-fox, attacking him if he comes too near. Gradually she becomes more tolerant until they are eventually inseparable. They groom one another, sleep together and play.

Play seems to be initiated by the dog-fox. He will approach the vixen and the pair will greet each other with mouths open. Then the dog rushes off and, if he is not followed, he returns to repeat the greeting. It may even be necessary to persuade the vixen to her feet by prodding and nipping. Eventually the games start. The dog-fox stands in front of the vixen and bumps her with his hindquarters, or they face one another and spar with open mouths. More vigorous games include 'King-of-the-castle', in which one fox takes up position on a log or other point of vantage and the other runs around it and charges in at intervals to spar. The most spectacular, and least seen, game is the dance, in which the pair rear up, facing each other, with forepaws on each other's shoulders, ears laid back and mouths agape. At the same time, they utter a chattering call, very similar to that of a flock of jackdaws. Both this cackling and the tail wagging which is a feature of this game are signs of a submissive attitude.

As a preliminary to mating, the dog-fox stands in front of the vixen and passes his tail over her shoulders. This may be to mark her with scent from his tail gland, which is particularly active around the mating season. Copulation ends with the 'mating tie', as in domestic dogs. The bulbus glandis at the base of the penis swells and prevents withdrawal for several minutes—the record time for fox copulation is forty minutes, but twenty minutes is more usual. While waiting for detumescence, the dog-fox dismounts and stands beside the vixen or faces in the opposite direction. The function of the mating tie is thought to be to ensure that the male's sperms have a good chance of effecting fertilization and are not supplanted by those from a mating with a second male. The advantage of a mating tie can be appreciated in sociable, promiscuous species but, if the fox is monogamous, it would seem superfluous.

That foxes are not always monogamous is shown by German observations where several dog-foxes kept company with one vixen (Schmook, 1960), but the dominant dog-fox performed most of the mating. Only later did the subordinate dog-foxes gain access to the vixen, and probably when she was no longer in full oestrus. The opposite to this polyandry has been described in the wild by Macdonald who found one dog associating with four vixens, who may have been related.

Fertilization occurs in January or February in England, and nine weeks later in northern Scandinavia. The cubs are born after fifty-two or -three days' gestation. The usual litter is of four to six cubs, but there is a record of seventeen from North America. Before birth, the vixen finds a dry place in which to rear the family. This may be in an earth den, under boulders or the exposed roots of a tree, in a badger sett or under a garden shed. In fox hunting districts, artificial earths of stone may be constructed for the vixen's benefit. This is not altruism: it is easier to dig out the cubs. No nest is made and the cubs lie on the bare floor. The birth is heralded by inactivity on the vixen's part and she will only make short hunting excursions until the cubs are a fortnight old. Until they are three weeks old the cubs cannot regulate the body temperature and they keep warm by huddling together. Later the vixen no longer sleeps in the earth but lies up nearby.

A fox cub is born with eyes and ears closed. The eyes open at eleven to fourteen days and the ears become erect at four weeks. At first the body is covered with short, greyish hair but this turns to dark choco-late colour by two weeks. Black eyestreaks develop at four weeks and at eight weeks the original woolly coat begins to be covered by longer guard hairs growing through it. The milk teeth begin to erupt at about four weeks and by eight weeks the snout has lengthened so that the cub now looks like a young fox, rather than an appealing puppy.

When four weeks old, the cubs receive their first solid food in the form of meat regurgitated by the vixen. Later, she carries food in her mouth and the cubs have to deal with it themselves. A vixen returning to her earth in the Highlands of Scotland was disturbed into dropping her prey. It consisted of a young rook, with five pheasant chicks and a mouse tucked under its wings! Weaning is complete by eight weeks. The first tentative steps above ground will have been made two weeks earlier but even now the cubs do not venture farther than a few metres from the safety of the earth. A bark of warning from the vixen sends them scurrying below. In areas of abundant cover the earth is aban-doned when the cubs are about ten weeks old and they then live on the surface. If disturbed, the vixen will move the cubs to another den, sometimes over a kilometre away.

The role of the father during the nursery period is uncertain. There are plenty of records of a dog-fox being found in the earth with the vixen and the cubs but it is quite possible that he is doing no more than paying a visit. However, Eric Ashby has filmed a dog-fox playing with the vixen and cubs. The dog-fox may feed the vixen during the first fortnight of the cubs' life or he may cache food nearby so she does not have to move far from the cubs to find it.

Two dog-foxes kept by Jane Burton have been paragons of paternal solicitude. Normally selfish in the extreme at feeding-time, their behaviour changed radically the moment the cubs were born. They would take the food to the vixen and feed only when she was satisfied, and later they brought food to the cubs. One vixen died on the point of giving birth. Her mate forced his way out of the pen, something never managed previously, killed a bantam and brought it back, pressing it against the body of the dead vixen.

If captive dog-foxes are such good parents it would be surprising if there was no sign of paternal care in the wild. There are at least two reports of a cub being discovered in the care of the father a week or so after an earth had been dug up, one cub having been overlooked when the vixen and the cubs were killed. A rare variant on the parental scheme is the joining of two vixens to rear their cubs in one earth. Possibly the seventeen American cubs mentioned earlier is an example of this. How the duties are shared is not known.

Serious training for adulthood starts when the cubs are twelve weeks old. Apart from boisterous play around the entrance of the earth the cubs spend much time investigating their surroundings. Every object is examined with the senses focused and finally it is tested for edibility. The vixen leads her cubs on hunting forays but she does not teach them to hunt. The nearest thing to instruction comes when the vixen holds food in her mouth and, moving her head from side to side and up and down, makes the cubs jump to get it. Hunting techniques are inborn but knowledge of what to hunt comes by trial and error aided by a lively sense of curiosity. There is, however, an indication that preference for a certain diet is passed as a tradition from one generation to the next, so that a poultry stealing vixen produces poultry stealing juveniles. She has not trained them to seek her favourite food, rather the cubs will have acquired a taste for the food that was brought to them in their infancy.

At the end of the summer, when four months old, the cubs become independent. Judging by the behaviour of captive animals, they are forced to move by the increasing irritability of the vixen. Female juveniles tend to stay near the parental home. Gwyn Lloyd found that Welsh vixens move no more than 2·2 kilometres on average but young dog-foxes move farther afield, on average 13·4 kilometres in sheep country where foxes live at low density.

According to Tembrock, foxes have a repertoire of twenty-eight calls. The best known are the eerie, hollow 'vixen's scream' and a harsh, short bark. The significance of the many screams, barks, yaps and howls are not well known and observers of fox behaviour disagree as to the situations when calls are used, as well as to their meaning. It may well be that some calls are used in a variety of situations, as when the domestic dog barks defiance, or to be 'let in' or just from excitement. The so-called vixen's scream, which is heard during the first half of the year, is known to be uttered by dog-foxes as well, and there is controversy whether or not screaming occurs during copulation.

Newborn fox cubs whimper when they are cold and hungry, but when they are capable of maintaining their own body temperature the whimpering turns to barks which are used as contact calls. Warning from a parent of danger is a single bark and a growl calls the cubs to food; panting and growling is a greeting between parents and cubs. The growl is also used as a contact call between adults at mating time. Adults threaten with loud whines and explosive hisses, and show submission by yapping plaintively and cackling.

The fox is also well equipped for scent communication through the possession of a number of scent producing glands. How this form of communication operates is largely a matter of conjecture based on what can be deduced from observations of marking and scenting behaviour. It is, for instance, typical of foxes, as well as cats and wolves, to sniff each other's lips when they meet. This suggests that the circumoral glands on the lips indicate identity or status of the individual, but there is no evidence as to the use of the foot gland, a group of sweat glands which open up in front of the large plantar pad on each foot. It can only be conjectured that the fox uses this scent to retrace its steps in unfamiliar territory.

Underneath the root of the tail and lying each side of the anus, there are two anal or subcaudal glands. On the dorsal surface of the tail some 4 to 5 centimetres from its root and masked by a tuft of dark hair, there is the tail gland, sometimes called the violet gland because its secretions are said to smell like violets. Both these glands show a peak of secretory activity around the time of mating and could be used by both dog-fox and vixen for marking in sexual and territorial contexts. Secretions of the anal glands are also deposited on stones, tufts of grass and other objects, perhaps as territorial markers. The most important scent-mark is the fox's urine. Urine is deposited at intervals in the manner of domestic dogs and David Macdonald's captive vixen marked 150 times in one hour, depositing a few drops of urine on certain regularly used places, which are very easily recognized even by the human nose. The faeces are deposited in conspicuous places but there is as yet no positive identification that they also act as scent marks.

The density of fox populations is very variable. In the Scottish hills it varies from one fox per 28 hectares to one per 4,000 hectares. How the country is divided among the foxes must depend to some extent on how many foxes are removed by hunters and whether they take a preponderance of dogs or vixens. Each fox has a home range where it has several lying-up places and a number of regular trails. It will spend a few days in one part of the range, then shift to a new zone. Ranges of neighbouring foxes may overlap but, during the breeding season, the dog-foxes become territorial. Each defends a part of its range against other dogs. The boundaries are established by scent marking and perhaps by barking. When neighbours encounter one another fights are rare and the contestants merely scream defiance. In North America, Vincent (1958) watched foxes coming to bait where a hierarchy was established at what was presumably neutral ground between their individual territories.

The vixens have a separate society within the males' territories. They scent-mark but the aim, according to Roger Burrows, is to establish a social hierarchy rather than to parcel out the land into territories. They thereby live in a loose community, perhaps of an old vixen with adult daughters who have not dispersed. Depending on population density, several vixens may live in the territory of one dog with polygamy ensuing

ARCTIC FOX
(Alopex lagopus)

Several carnivores penetrate the tundra of northern Eurasia and North America but, apart from the polar bear, no other carnivore is virtually confined to the tundra nor so well adapted to life in cold climates as the Arctic fox. It is superficially similar to the red fox but smaller. The fur is much longer, especially in winter, and the muzzle and ears are short, giving the fox an almost cat-like appearance. There are two colour phases. In one, the summer coat is a smoky grey-brown with white underparts and the winter coat is either pure white or slightly cream. During the moult the fox becomes oddly skewbald as patches of fur change colour. The 'blue fox' is a colour variety which exists naturally in small numbers and is specially bred on fur farms. The uniformly bluish-grey is retained throughout the year except for a whitening on the tips of the hair. The body length is 50–65 centimetres with a tail of 28–33 centimetres and a height at the shoulder of 30 centimetres. Weight ranges from 4·5 to 8 kilograms.

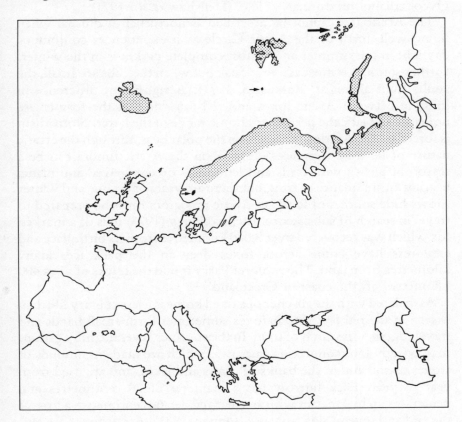

The range of the Arctic fox lies mainly within the Arctic Circle but thrusts southwards around the shores of the Bering Sea and Hudson Bay, in Greenland, Iceland and Scandinavia. The European distribution is confined to Iceland, northern Finland and Russia and the mountains of western Norway. Through its ability to travel over the frozen sea, the Arctic fox has also been able to colonize Svalbard, Bear Island, Franz Josef Land and Novaya Zemlya. During the winter, some Arctic foxes migrate southwards, penetrating the treeline, and temporarily colonize southern Norway, much of Sweden and Finland, but Braestrup (1941) knew of only one instance of Arctic foxes living permanently among trees. Those that stay behind on the tundra do not hibernate but remain active. Arctic foxes have been seen hunting in temperatures of −45° C. By experiment it has been shown that an Arctic fox can maintain a temperature of only one degree below freezing in the pad of its paw when it is immersed in a water/glycol/ethanol mixture at −35° C (Henshaw et al., 1971).

The Arctic fox cannot be described as nocturnal or diurnal when living well north of the Arctic Circle as it experiences continuous daylight in the summer and almost complete darkness in the winter. Farther south, it is most active at dusk but when the cubs are small, the adults hunt all night (Kaikusalo, 1971). A significant difference in lifestyle between Arctic foxes and red foxes is that the former are nomadic and lack the permanent home range of the latter. Nomadism is forced on the Arctic fox, as it is on the polar bear, through the erratic nature of its food supplies. Summer on the Arctic tundra can be a season of plenty with birds and lemmings breeding well and plants bearing an abundance of fruit, but there are years of famine and winter snows hide sources of food, so Arctic predators must be prepared to travel in search of subsistence. Richard Perry (1973) tells of a marked fox which was recovered over 9,000 kilometres from its birthplace and explorers have come across foxes deep in the pack ice, many kilometres from land. The explorer Parry found the tracks of a fox 400 kilometres off the coast of Greenland.

Associated with the absence of a fixed home is a less solitary life than is seen in the red fox. Arctic foxes sometimes form small bands and travel together in search of food. In the summer several family groups may occupy interconnecting burrows, which are made in mounds of stones or soil and in the banks of rivers and lakes, and are used from year to year. Each burrow has a diameter of 20 centimetres and penetrates up to 4 metres with an enlarged den 60 centimetres across at the end and several side passages (Ognev, 1931). Arctic foxes are also less wary than red foxes, a trait which is both a pleasure and a penance for visitors to the Arctic. Arctic foxes will fearlessly enter a camp and can be hand fed but they will also help themselves to anything edible within reach.

In the European part of its range, the Arctic fox's main food is voles and lemmings and fox numbers fluctuate with the cyclic rise and fall of small rodent populations (Corbet, 1966). Braestrup (1941) distinguishes between 'lemming foxes' whose breeding fluctuates with the lemming cycle and 'coastal foxes' which subsist on seabirds and breed regularly. Arctic foxes hunt rodents in much the same way as do red foxes. They listen for the sounds of movement under the snow or tundra vegetation and pounce, bringing both paws down simultaneously. There are no lemmings or voles in Iceland and there the foxes have to rely more on birds and what can be found on the seashore in the way of cast-up seal and whale carcases, shellfish and sea urchins, even beach flies.

The summer is a time of plenty for the Arctic fox. Not only are the rodents breeding, but many birds invade the tundra to rear their broods. Predation by Arctic foxes has moulded the nesting habits of these birds and they have evolved strategies to safeguard their offspring. Some choose a safe, inaccessible nesting place. Gulls and geese often nest on cliff ledges; terns nest on small islands. Skuas harry intruding foxes with aggressive dive-bombing tactics and waders lure foxes from the nest with distraction displays—making themselves conspicuous with the so-called 'broken-wing' and 'rodent run' displays. Eider ducks sit tight on their nests and rely on superb camouflage and immobility to protect them, but nevertheless, many nests fall prey to foxes. In coastal regions there are huge colonies of auks, known as looneries, which are a great attraction for foxes. Guillemots on cliff ledges and puffins in their burrows are safe but the prize pickings come when the young birds, especially guillemots, leave the nest. Guillemot fledglings leave the cliff ledge before their flight feathers are fully grown. They flutter down to the sea and swim away from the shore but some fall short and crash on to the rocks where the foxes can pick them up. Foxes may also be able to rob the nests of little auks and black guillemots which are found under the boulders of the talus slopes at the bottom of sea cliffs.

Other summer foods include berries and, rarely, fish. Ognev (1931) gives an eyewitness account of Arctic foxes catching loaches by crashing through thin ice and reappearing with fish in their mouths. When travelling over the pack ice, the foxes subsist on birds and fish caught in tidecracks and leads or they follow polar bears to feed on their leftovers. Unless famished, a polar bear frequently leaves much of a seal carcase to its attendant foxes. In the spring the foxes join the bears in digging ringed seal pups out of their dens under the snow.

When there is a surplus of food, caches are made. Prey is brought back to the den or stored in crevices for use in early spring when food supplies are at a minimum. Alwin Pedersen (1966) found a cache containing thirty-six little auks, each with its head bitten off, two guillemots, four snow buntings and a large quantity of auks' eggs. The corpses were buried neatly with their tails facing the same direction.

One of my Arctic camps was troubled by a fox which took to stealing and caching bars of soap.

The breeding season starts in April. Males become territorial and establish their position with urine marks, with the same 'leg-cock' as a domestic dog, as does the female when in oestrus. Faeces are also probably used as scent-marks (Kleiman, 1966). The tail gland is well developed, and there are large glands giving a 'foxy odour' on the paws but their function is not known. Pairing is for life and the sexes maintain contact with a three- to five-note bark. After a gestation of about fifty-three days, a litter of six or seven cubs is born in the den among rocks or in a burrow. A food shortage results in a smaller litter through prenatal mortality but there are two litters per year. The female mates again three weeks after the birth of the first litter and the second litter is born in July or August. Two females may share a den and care for each other's cubs.

The cubs open their eyes at two weeks and first leave the den, with their mother in attendance, at three weeks. They are weaned at six weeks when both parents bring meat back to the den.

RACCOON DOG
(Nyctereutes procyonoides)

The raccoon dog is a native of the Orient which has been introduced to Eastern Europe as a valuable fur bearer. It can be described as a cross between a red fox and a raccoon, being about the size of a fox but with a short, blunt muzzle, a stouter body and shorter legs. The tail is rather short and bushy. Head and body length is 55–65 centimetres and tail length 15–17·5 centimetres. The name is well earned from the patch of black around and under each eye, sometimes meeting over the muzzle, which resembles the 'robber mask' of the raccoon, together with a tuft of long hair on each cheek. The fur is yellowish-brown with a dark collar over the shoulders, and dark legs and tip of the tail.

The native home of the raccoon dog is the warmer, forested regions of eastern Asia from the Amur River to northern Vietnam. It was introduced to European Russia in 1927 as a source of pelts and it has

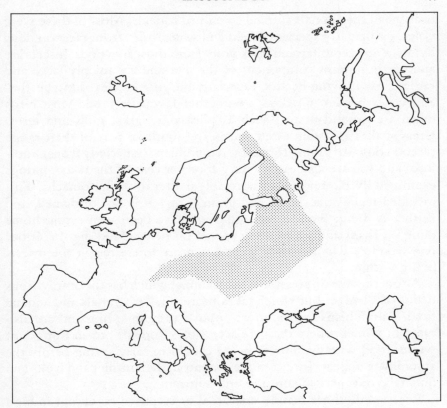

spread rapidly westwards. Raccoon dogs thrive especially around
Moscow, Leningrad, Kalinin and Smolensk and in the Pripet Marshes.
They have moved westward and are now found in Poland, Hungary,
Romania and Czechoslovakia. Germany and Switzerland have been
invaded and, to the north, raccoon dogs are established in Finland and
Sweden and are moving into Norway.

In the Far East, raccoon dogs inhabit open broad-leaved forests with
dense ground cover, particularly alongside water. The new European
habitat is similar with raccoon dogs shunning coniferous forests,
steppes and marshes.

The raccoon dog has a more varied diet than most of the Carnivora,
taking more vegetable food and hunting the smaller animals. It takes
more warm-blooded prey than the badger but it shows the same
trends away from flesh-eating in the reduced canines and carnassials,
flattened molars and relatively long intestine. Bannikov (1964) investigated the diet of raccoon dogs living in Russia. The stomachs of

nearly half the animals studied contained rodents. Most of these were voles, particularly species of field and water voles from raccoon dogs living in open country and bank voles from those in woods. Insects are another important component of the diet and are mainly large and rather slow moving beetles, especially burying beetles, dung beetles, and water beetles. Crickets are another favourite food. Most birds taken are ground-nesters such as pheasants, larks, gulls and terns. Grass snakes are often eaten and in the southern part of their range raccoon dogs eat young tortoises. Amphibians (especially frogs), molluscs and fish are also eaten. Sixty-four per cent of the raccoon dogs examined by Bannikov had vegetable matter in their stomachs. This included fruit, nuts, bulbs and grain. Carrion is not disdained, and neither is dung: Finnish raccoon dogs have been seen eating horse dung on roads and sheep dung spread on fields (Mikkola, 1974) but one wonders whether they were not more interested in the insects living within.

Altogether the impression is of an animal which has the omnivorous habits of a badger but which takes more small mammals and aquatic food. Amphibians are the most popular prey in spring and are also eaten in winter where they are accessible. Rodents are an important winter food with even more being caught in early spring before the amphibians appear. Birds replace rodents in the summer and fruits and nuts take over in late summer and autumn.

Raccoon dogs are mainly nocturnal or crepuscular, except in high summer when nights are short. During the day they lie up in beds of reeds or rushes, in hollow trees or scrub, and under rocks or over-hanging banks. Abandoned fox and badger holes are used and sometimes a raccoon dog excavates its own burrow in a slope, often facing east. The main tunnel is about 2 metres long and 25 centimetres in diameter and ends in a simple chamber. There are usually one or two entrances but there can be up to five. Dung pits are located around the entrance (Bannikov, 1964).

When the winter snow cover starts to build up raccoons rapidly become inactive and fall into a deep sleep with the metabolism dropping by 20 per cent (Bannikov, 1964). It is always said that the raccoon dog is the only member of the dog family to hibernate but this is merely a denning-up and relative inactivity, not true hibernation. This prolonged sleeping must be made possible by the raccoon dog's omnivorous habits, which allow it to gorge on fruits in the autumn to lay up a store of fat for winter in the manner of the similarly part-vegetarian bears and badgers. In Finland, raccoon dogs disappear between November and April but they reappear on mild days and remain active throughout warm winters (Mikkola, 1974).

The rut starts when activity is resumed in spring, although pairs will have formed in the previous autumn. The female is on heat for only a very short time, from a few hours to a few days, but there is a second heat three weeks later, even if the raccoon dog has become pregnant during the first (Bannikov, 1964). During the heat period, the female scent marks with urine by fully cocking her leg like a dog, an unusual habit (Ewer, 1968). The litter is born in April or May, after a gestation of 51–70 days. There are usually six or seven cubs but up to sixteen have been recorded. They are blind at first and have short dark fur. Weaning is accomplished at 45–60 days but they will have started to eat insects and frogs some weeks earlier. The father plays an important part in bringing food to the cubs and the family sleeps together in the same burrow.

WOLF
(Canis lupus)

The wolf was once the ecological counterpart of prehistoric man: a sociable animal eating a variety of foods and often co-operating to hunt large prey. While the human population still consisted of small, widely scattered nomadic bands there can have been little competition, but the advent of animal husbandry, when man started to keep large herds of vulnerable herbivores, brought the two into conflict and man, himself, felt threatened. The Saxon king Edgar commuted the sentences of convicts in return for their killing a certain number of wolves. In Highland Scotland, as elsewhere, spittals (a name corrupted from 'hospital') were established in wild country as safe resting places for benighted travellers and, for the last security, burial places were established on small islands (Darling and Boyd, 1964). Nevertheless, the wolf's influence on human history has not been wholly negative. The wolf is the progenitor of the domestic dog, Man's Best Friend, and was fabled foster-mother to the co-founders of Rome.

The wolf can be readily described as resembling a large dog, an

Alsatian for instance, but it is more powerfully built. The ears are smaller and more erect, the tail is rather long. The face is broader due to the large cheek muscles and the teeth are larger. The overall appearance of bulk is increased by the thick coat which forms a mane around the muscular neck. In all, the wolf is a stronger, more powerful carnivore than any domestic dog. The body length is 110–140 centimetres, with a tail of 30–40 centimetres, and the height at the shoulder is 70–80 centimetres. The weight range is 25–50 kilograms. The colour of the coat is very variable. In the far north the fur is almost white and much longer, but farther south it darkens to tawny or brown with a black shading on the back. In some populations there is a black stripe running down the forelegs.

The wolf's distribution is holarctic, encompassing the cooler regions of both Eurasia and America. At one time it ranged over most of North America, including northern Mexico and western Greenland, all Europe and the bulk of Asia, except the Orient. It is now

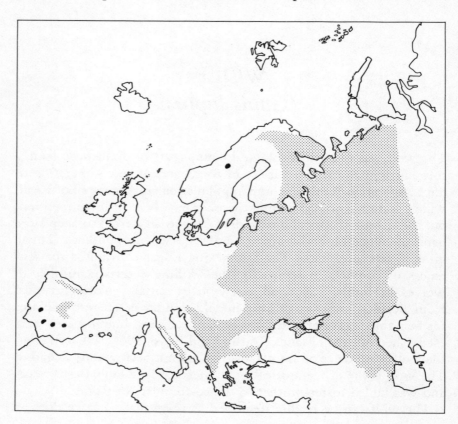

extinct in many parts and very rare in many others. The last British wolf was killed in 1743 in the Scottish Highlands, but small populations linger in Spain and Portugal, Italy, northern Greece and the Balkans, and Fenno-Scandia. Wolves are faring better behind the Iron Curtain, according to Pulliainen (1965), because of changes in human society. The old estates have been taken over by the State and hunting for sport is discontinued. As a result wolves are spreading westwards not only into Finland but across Germany. Wolves are found in forests and on the tundra and steppe. Arid deserts are said to be shunned but Wilfred Thesiger (1959) found a wolf when crossing the Empty Quarter of the Arabian desert. Presumably, however, this wolf had wandered into the desert and was not a permanent resident.

One of the traditional images of the wolf is of a pack of a score or more in hot pursuit of a horse-drawn sleigh hurtling through a snow-carpeted forest. In the main, however, man-hunting is not a characteristic trait (although the sleigh-horses are a different matter) and wolves do not live in ravaging packs. The wolf is undoubtedly sociable in comparison with most other carnivores but its relationships depend on circumstances. It is not altogether clear from studies of wild wolves how their social life is organized. Large packs may be formed in the harshest climatic conditions but the usual group size is much smaller and consists of a family party. In Finland, Pulliainen has found that the average pack consists of no more than three or four wolves and solitary wolves are frequently seen.

The basis of wolf social life is the family: mother and her litter of cubs. When the cubs begin to take an active part in hunting at the age of seven or eight months, they do not go their separate ways, as do most carnivores, but stay with their mother. She is the leader and takes the initiative in communal hunting, although she may well not be physically in the lead when the pack is travelling. If, however, a male is consorting with the female, he will be leader. The pack attracts other wolves which have spent the summer living alone and sometimes two families join forces. To give the lie to stories of huge wolf-packs chasing travellers, a pack of over twenty wolves is rare and even a pack of ten is a large pack.

Individual relationships between members of the pack have been studied in captivity but even where wolves have a large compound in which to roam, the situation is artificial because individuals are not able to wander off to live alone. In these limiting circumstances, two hierarchies of rank form, one of males and one of females. Position in the hierarchy is first established by fighting and then maintained by

ritual, making use of signals which indicate aggression and submission. Superiors parade their dominance by looking aggressive.
They approach subordinates with an erect carriage, the ears are
pricked, the tail is raised and the lips may be parted in a snarl to show
off the teeth. To avert trouble, the subordinate becomes self-effacing.
It creeps up on bent legs, with the tail lowered and the ears flattened
and licks its superior's muzzle. If this does not allay the menace, it rolls
on its back and presents its vulnerable throat in complete submission.
Members of the pack also communicate by howling, a doleful, drawn
out call which keeps them in contact. Like the red fox, the wolf has a
gland on the upper side of the tail, but the wolf's is poorly developed.
Wolves have only rarely been seen to sniff each other's tail gland and
its role is not known (Fox, 1971a).

Each pack has a home range which it traverses throughout the
winter by following established trails. Home ranges of American
wolves vary from 315 to 1,250 square kilometres for a pack or 25 to
268 square kilometres per wolf (data from Ewer, 1973). Very often the
hunting trails follow the migration routes of prey animals. Distances
covered range from 20 kilometres or so a day in forests up to 200
kilometres when running through open country on hard snow. The
pack breaks up in the spring and the population becomes sedentary as
the females become tied down by their maternal duties (Pulliainen,
1965 and Ognev, 1931).

Courtship starts in January and continues through to March. There
may be considerable fighting as young animals are ousted from the
pack and in captive situations, perhaps also in the wild, only the
dominant male and female mate. Gestation lasts 60–63 days but before
giving birth, the female wolf selects a suitable den. She may dig a short
burrow in sandy soil, enlarge the burrow of badger, fox or marmot, or
lie up under boulders and under thick bushes. The litter usually varies
from one to seven cubs, being generally fewer in a female's first litter.
The record stands at thirteen but, as with the red fox, there is also the
suspicion that such extreme records are of two litters combined. The
cubs are blind at birth and lie huddled together like domestic puppies.
Their mother spends much of her time with them at this stage and
hunts in the near vicinity of the den. Their eyes open in 10–12 days and
by three weeks the cubs play around the entrance of the den.

The male wolf plays an important part in rearing the family. While a
female is capable of raising a litter on her own, she benefits from food
brought by her mate when she is tied to the den by newborn cubs.
Later, the male helps to feed the growing cubs, a trait which must

greatly improve chances of survival in large litters. Both parents bring them prey, either carried in the mouth or regurgitated from the stomach. An interesting observation both of paternal solicitude and the intelligence of wolves was of a captive male wolf which herded his cubs indoors when the keeper hosed down the pen outside. The wolf shut the door behind the cubs by tugging a pulley rope and later opened it with his muzzle and paw (Fox, 1971b). As the cubs grow older they wander farther. When five or six months old their mother's labours are greatly reduced because they can follow her and feed at the site of the kill and, two months later, they are learning to catch their own prey. The scene is now set for the re-establishment of the pack. Heralded by an outburst of howling as the neighbourhood's wolves make contact, young wolves, the previous year's cubs who were driven away in the spring, give up their solitary ways and join the packs.

The wolf's diet is varied, but where available the main food is hoofed animals. The expression 'to wolf one's food' is well-deserved for a wolf can bolt up to one-fifth of its own weight at a sitting and everything is digested except hair, hoofs, horns and feathers. Such voracity is particularly necessary in winter when a wolf may go several days without making a kill. Foods range from elk, reindeer and cattle to cats, mice and fruit, and Mediterranean wolves plunder grapevines. Carrion is not disdained and scavenging around human habitation can be an important source of food. In Finland, sheep and cattle are the main summer food but when in winter the cattle are confined in or near the farms, the wolves must prey on wild animals. Wolves are the most important predator of reindeer in northern Finland and, in some parts of that country, sheep farmers have been forced out of business by their depredations.

Hunting methods have been studied in North America by David Mech (1970); he observed that large animals were discovered in three ways. They are most often found by accident as the wolves quarter their range. In other instances, the wolves discover trails and follow them, or they may scent them from a distance. Scenting is usually accomplished when the prey is within 300 metres. However, on one occasion Mech observed a cow moose, with twin calves, which was scented from 2·5 kilometres. Opinions differ as to whether wolves selectively prey on young, old or diseased animals, or whether they kill at random; the choice to a great extent must depend on the situation. A wolf, with its broad paws, has the advantage over a narrow-hoofed deer in the snow and can kill as it pleases, but, where the prey animals can make a good escape, the wolf has to select the slowest. On firm ground a healthy deer can outrun wolves, which give up the chase as soon as it appears hopeless. Mech found that a healthy moose (the same animal as the European elk) can survive an onslaught by wolves if it stands and defends itself with its hoofs, but if the wolves can force it to flee, they increase their chances of overwhelming it. When the wolves catch up with their quarry they tear it down by leaping up to bite and tear at its flanks and legs. The pack then feeds amicably, all animals sharing the spoil without quarrelling.

Where hoofed animals are available, small animals are of less importance as food and serve only to make up any deficiencies left by the main prey. If there are no hoofed animals, wolves have to subsist on small fry such as hares, rodents, birds and occasionally reptiles and amphibians, as well as the larger foxes, badgers and dogs. Such small animals are either run down or stalked and pounced upon. Bears, lynx and wolverine wisely keep clear of wolves as they may become their prey (Pulliainen, 1965). Only wild boar seem to be immune from attack and are unconcerned by the proximity of wolves (Ognev, 1931).

GOLDEN JACKAL
(Canis aureus)

The golden or Indian jackal has a geographical range which covers northern and eastern Africa and southern Asia. Its inclusion in the list of European carnivores is due to its range extending well into southeast Europe. Unlike other principally Asian carnivores, such as the tiger and the hyaena, the golden jackal is well entrenched in southeastern Europe. It is found in Russia and the Balkan countries, and occasionally turns up in Hungary. In appearance, the golden jackal looks like a long-muzzled slender dog, with large ears, long legs and a bushy tail. The coat is dirty yellow or tawny with an admixture of black and brown hairs on the back. The tail is reddish-brown with a rounded black tip. Combined head and body length is 84–105 centimetres, with the tail an extra 20–24 centimetres. Height at the shoulder is 50 centimetres and the weight 10–15 kilograms.

In contrast to the other species of jackal living in Africa, little is

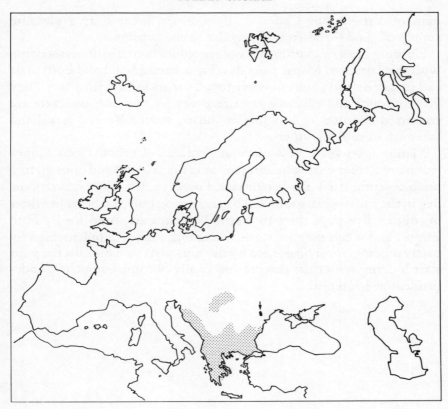

known about the habits of the golden jackal and its study has been
almost entirely neglected in Europe. Its European home is the steppes
and semi-deserts and other sparsely vegetated semi-open country
where there are thickets and scrub to give good cover. Reedbeds are
popular but woodland seems to be shunned. The semi-deserts of the
Caucasus are principally a summer home as the jackals cannot survive
their harsh winters and deep snow (Vereschagin, 1959).

Like its cousin, the red fox, the jackal has an omnivorous diet. It is
famous as a scavenger and is appreciated by the Balkan peasants for
clearing refuse, and now the search for scraps frequently brings jackals
into towns and cities. In addition to scraps and carrion, the jackal hunts
for small animals. Rodents are captured by the same sudden pounce as
is used by foxes. Birds and their nests are regularly attacked, and in
some parts frequent losses among poultry have led to the persecution
of jackals. Larger prey, such as sheep, are occasionally killed by two or
more jackals working in concert. Insects and molluscs are a common

item of diet—in the Caucasus, in summer, locusts are a plentiful supply of food—and fruit is popular in the autumn.

Hunting in a group probably occurs only when cubs are associating with their parents. Mated pairs develop a strong bond and both male and female defend a common territory by marking it with urine. They keep in touch and advertise their territory by howling and there are concerted choruses of howling at sunset, started by one jackal and taken up in turn by others.

Mating takes place in January or February (Corbet, 1966). Copulation involves the same 'mating tie' as in foxes (p.40) and once mating has been completed, the female jackal makes a den. She excavates the den in the soil or enlarges an old badger or fox hole and bears her litter of four or five pups there in April. The pups are cared for by both parents and when they are three weeks old, milk is supplemented by partly digested meat disgorged by the adults. At two months the pups start hunting with their parents and finally become completely independent at six months.

The Bear Family
(Ursidae)

The bears are large mammals which have evolved away from a predatory way of life. They are heavily built and rather slow moving. The hindfeet are plantigrade. Each foot has five toes bearing long claws, and bears are adapted for moving in rough, mountain country or in forests. Small bears are good tree-climbers. The tail is small, as are the rounded ears and the eyes. The fur is coarse and dense.

Bears are solitary. They feed on a variety of vegetable foods and small animals and occasionally kill large animals such as deer. The teeth are blunt and adapted for chewing tough plant food. The exception is the polar bear which has returned to hunting and has the long canines and cusped molars of a meat eater. Contrary to popular myth, bears do not hibernate but lie up in winter dens. The cubs are born after delayed implantation, while they are very small and about one three-hundred and fiftieth of the adult weight.

The eight species of bears are virtually confined to the northern hemisphere but the Malayan sun bear extends through south-east Asia into Sumatra and Borneo, and the spectacled bear lives in the mountains of South America as far south as Bolivia. The giant panda is now generally considered to be a bear rather than a member of the family Procyonidae but this is not accepted by all zoologists. Continental zoologists have traditionally assigned the giant panda to the bears, while in English-speaking countries it is placed in the Procyonidae. The problem is that, in adapting to a diet of bamboos, the giant panda has developed features differing from either group of possible relatives. European bears number the brown bear, which is conspecific with the American grizzly bear, and the maritime polar bear.

BROWN BEAR
(Ursos arctos)

Once widespread over the continent of Europe, the brown bear has been driven back by the spread of civilization. Although it has been persecuted because of its threat to the safety of men and their livestock and also because its fur is prized, as formerly was its flesh and fat, destruction of habitat is the overwhelming cause of the bear's disappearance.

The brown bear has, in the past, been classified into half a dozen species and as many subspecies but the distinctions were based mainly on variations in colour and size. J. G. Millais (1904) wrote: 'No terrestrial mammal varies so greatly, both in size and pelage, as this animal. Between brown bears killed in eastern Norway and those of western Sweden there is perceptible difference in colour, whilst in the bears of Russia, especially those of the eastern districts, there is a further much greater difference in size.' The colour range is from cream to dark brown or black, sometimes with darker legs. Juveniles have a light collar and the coat may darken through the year. Body length is from 170–250 centimetres in males, with females smaller. The tail is so short that the brown bear appears tailless. Males weigh 105–265 kilograms and stand around 100 centimetres at the shoulder. The stance is plantigrade on the hindfeet, digitigrade on the forefeet, the feet being broad and the weight taken on the cushion-like pads. Each paw has five toes armed with long, non-retractile claws.

If those zoologists who think that the grizzly bear and Kodiak and Kenai bears of Alaska are no more than subspecies of brown bear are right, the brown bear has joined the ranks of animals with a holarctic (combined North American and Eurasian) distribution. In the Old World, brown bears are found from Scandinavia and the Pyrénées to the eastern tip of Siberia. The original range extended westwards to the British Isles but bears have largely disappeared from western Europe. They vanished from Denmark five thousand years ago and from the British Isles in the tenth century. Elsewhere, they lingered; the last bear in Germany was seen in 1836 and the last in Switzerland in 1914 although this was a visitor and native bears had died out earlier (Curry-Lindahl, 1972).

The European bear was equally at home in deciduous or coniferous forests, provided there was a rich undergrowth. Now the deciduous

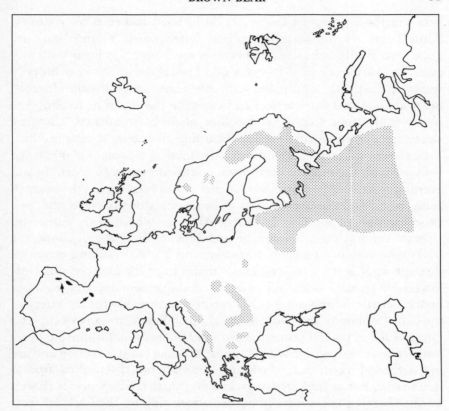

forests have largely disappeared, its last strongholds are the conifer forests of wild mountain regions and the marshes and bogs of eastern Europe. The largest populations are in Russia and the Carpathians of Romania and Yugoslavia, with isolated pockets in the mountains of Scandinavia and Finland, the Cantabrian mountains and the Pyrénées of Spain and France, the Apennines and Dolomites of Italy and the Balkan ranges of Yugoslavia, Albania, Greece and Bulgaria. There are also bears in Poland, Austria, Czechoslovakia, Hungary and Estonia.

Although heavily built and not developed for speed, a bear can show a good pace over short distances. It normally walks on all fours, with the head swinging from side to side, but it can stand erect and shuffle on its hind legs for a few paces, a trait enhanced by training in the dancing bears of old. The upright stance is particularly used when the bear has been alerted and is trying to see or smell the source of disturbance. Its paws are seemingly clumsy, but they are well adapted for a number of tasks. As well as knocking down prey with them a bear can flip fish out of the water, hold food and turn over stones. Some bears are right-handed, others left-handed. Young bears are good tree climbers but adults become too heavy to climb with any ease. The claws are used to get a good purchase as the bear literally walks up the trunk. Compared with other carnivores, the shoulders of bears are adapted for traction and can take the strain of hauling the heavy body up a tree. The shoulder blade is broadened, allowing attachment of powerful muscles, including an outsize triceps muscle.

Bears are solitary except during the mating season and while the cubs are with the mother. Activity is confined mainly to morning and evening, particularly where human disturbance has forced the bears to shun the daylight hours. Groups of bears occasionally foregather at a concentration of food, as in North America where bears gather by rivers to catch salmon. There is no sociability in these aggregations; the bears take care not to come too close and a social ranking develops through aggressive conflicts. Large males have the highest rank, followed by females with old cubs, then single females and younger males. Females with young cubs keep away, probably for the safety of the cubs (Stonorov and Stokes, 1972). Each bear normally occupies a range of about 10–15 kilometres diameter but how neighbouring bears interact is not known. It would appear that the ranges overlap and are not defended territories. Marking points are not distributed around boundaries but at trail crossings and favourite feeding places (Ewer, 1973). Marking is performed by urination, ripping the bark off trees with teeth or claws and rubbing the neck, shoulders and chest against the trunk (Tschanz et al., 1970). Rubbing takes place all the year round but it is most intensive during the breeding season. Young bears are unwilling to go near rubbing trees used by their elders.

The teeth demonstrate that bears are carnivores which have become vegetarian. They have canine teeth and a full carnivore's dentition, but the teeth are broad and blunt. The molars are flat for grinding fibrous food and the shearing carnassials of typical meat-eaters are poorly developed. A wide variety of foods is taken, depending on habitat and season. Grasses, herbs and roots make up the bulk of the diet and turf is rolled up like a carpet to reveal not only the swollen roots but also worms and insects (Ognev, 1931). Ants' nests and beehives are raided and around Lake Baikal bears feed on abundant caddis larvae in spring (Bannikov, 1967). Fruits are an important source of food in autumn—Pyrénéan bears are known to eat bilberries, bearberries, currants, raspberries, pears and strawberries as well as acorns, beechmast and nuts. The list is amazingly extensive and includes such unlikely items, for a carnivore, as fungi, the stems of angelica and wild carrot, and the roots of ransoms, arum and trefoils (Couturier, 1954). In the far north, bears move on to the tundra for the berry crop. Grain crops are sometimes attacked. Brown bears revert to flesh-eating mainly when plant food is not available. Small animals—rodents, including dormice, frogs, the eggs and young of birds—are taken when the opportunity arises and some individuals develop a taste for meat. Late summer and autumn provide a cornucopia of fruits and young animals, but the spring is a hard time until the snow has melted. Russian bears spread on to the tundra to seek ground squirrels and the corpses of reindeer which died during the winter, while Italian bears subsist on early growing herbs, insects, bulbs and wild pears remaining from the previous autumn (Zunino and Herrero, 1972).

Bears have earned a bad name for attacking livestock but this is very much a trait confined to rogue individuals. Bears may even feed alongside cattle in fields, with neither species taking any notice of the other. Yet a bear can kill a calf or a pig with a single, tremendous swipe. Horses, sheep, wild boar, roe and red deer and elk are also attacked and cannibalism is not unknown (Ognev, 1931). Large prey which cannot be eaten at one sitting are cached by burial with moss and leaves. The bear stays near the cache until it is consumed and Mysterud (1973) considers that burial hides the carcase from ravens and other bears.

As with all large animals, the brown bear's relations with man are variable. In general it is not aggressive and usually flees if approached but attacks can occur without provocation, to the extent of a bear creeping up to an unsuspecting man. It has often been said that a special danger with bears is that they are rather silent and have expressionless faces, whereas a dog or cat shows its aggressive feelings with bared teeth and flattened ears. This is not exactly the case. Although not so obvious as a dog or cat, bears do communicate their feelings through the position of the ears, by opening the mouth and by body posture (Burghardt and Burghardt, 1972).

The autumn harvest of fruit is important in enabling the bear to lay down a good store of fat for the winter. Flesh-eating animals do not usually have such an abundance of easily obtained food in the autumn and cannot lay down such a thickness of fat. However, their food continues to be available through the winter while a vegetarian is hard-pressed, particularly if there is continuous snow cover and hard frosts. A vegetarian must rely on a good store of fat which can be conserved by reduced activity. Bears do not hibernate but they do show a considerable reduction in activity by denning up during the winter months. While in the den, the body temperature drops a few degrees and heart and breathing rates slow down, heart beat dropping from forty to ten per minute. There is disagreement as to how soundly a bear sleeps. Curry-Lindahl says that a bear is immediately active on disturbance but Ognev claims that a tree felled across a den will not flush the occupant.

Denning usually takes place at the first snow but a good supply of food causes a delay and in warmer places, as in the southern Caucasus, bears are abroad the year round. Around Moscow, denning starts in the second half of October and the bears emerge in April (Ognev, 1931). The warmer weather of the Pyrénées reduces denning from December to March (Couturier, 1954). The den is prepared in advance, under exposed roots, a fallen tree, a boulder, in a cave or a hollow tree, even in a wood ants' nest, preferably on a south-facing slope. It may be lined with mosses, leaves or twigs, or the bear may lie on bare ground. The entrance is blocked with branches and leaves. Sleep does not come immediately and the bear may emerge again in fine weather. Eventually it settles down on its side, its snout between its paws. If undisturbed, a den may be used for several years in succession.

The mating season lasts from May to July and again there is disagreement over details. Curry-Lindahl describes the brown bear as promiscuous while the Pyrénéan population is said to be not only monogamous but faithful from year to year. The she-bear is sexually active for 10–30 days, during which period the male follows her closely and smells her genitalia. By way of courtship he snorts frequently and licks or nibbles her head and neck. Most accounts describe mating in bears as following the usual mounting of quadrupeds *more canem* but Ognev says that the female lies on her back *more humanum*. This was believed by ancient writers. In a twelfth century Latin bestiary, translated by T. H. White (1954), the author, a monk plagiarizing much earlier texts, writes that bears 'do not make love like other quadrupeds, but, being joined in mutual embraces, they copulate in the human way'.

Delayed implantation takes place and the cubs are born in the winter den during January or February. The function of the delayed implantation is, according to Ewer (1973), twofold. It allows mating to take place before the autumn feeding and also makes birth sufficiently early for the cubs to leave the den with their mother and start weaning when food is plentiful. There are usually three cubs and there can be up to six, but this is rare. First litters, born when the mother is four years old, usually number one or two. The cubs are extremely small when born. Each is blind and helpless, less than 0·5 kilograms and around 22 centimetres long. The usual explanation of such premature birth is that the delivery does not exhaust the lethargic she-bear and she will not use up her precious winter food reserves in suckling large cubs. The placenta is eaten immediately and the cubs are licked and held against the mother's body in her arms.

The small size of the newborn cubs is reflected in the old belief that they were born as a shapeless mass of flesh and were 'licked into shape' by their mothers. The story must have arisen from someone witnessing the mother bear licking her cubs free of the birth membranes. The writer of the bestiary describes the newborn cubs as 'tiny pulps of a white colour, with no eyes', which are shaped by licking, and he goes on to explain that birth is premature and that she-bears hold their cubs to the chest and continue to lie up without food for three months after the birth—statements which are surprisingly accurate.

The cubs' eyes open when they are one month old and they start to follow the mother on feeding forays well before weaning at four months old. A certain amount of suckling continues and the cubs do not become fully independent until their second summer. As a consequence she-bears breed only once in two years.

POLAR BEAR
(Thalarctos maritimus)

The polar bear only narrowly qualifies for inclusion as a European animal. It is essentially a pelagic animal, spending much of its life on the Arctic pack ice, but many bears come ashore to den up in winter or to feed in the summer. Polar bears come ashore on continental Europe only as stragglers reaching the Arctic shores of Scandinavia and Russia but they occur regularly on offshore islands such as Svalbard, Franz Josef Land and Novaya Zemlya. The extension of many countries' territorial waters to 300 kilometres offshore may strengthen the polar bear's claim to European citizenship!

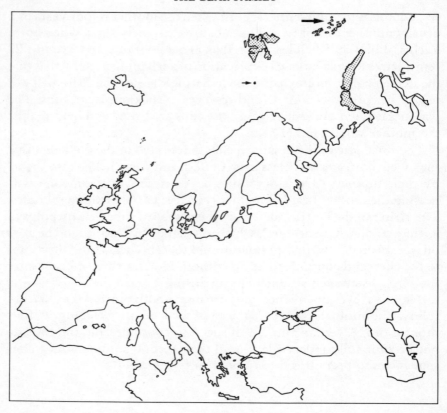

Although placed in a separate genus, *Thalarctos*, the polar bear is closely related to the brown bear, *Ursus*, having diverged from it in the last Ice Age, and interbreeding in zoos has resulted in healthy offspring. The unique features of the polar bear are a result of its adaptation to a particular, and hostile, environment. The polar bear is considerably larger than most brown bears. Males have a body length of 200–250 centimetres, stand up to 140 centimetres at the shoulder and weigh 400–450 kilograms. Females are smaller. The fur of older animals has a yellowish tinge that makes it look dirty. The neck and head are long, the ears and tail small and the snout is 'Roman' with a slight arch. The limbs are powerful and no visitor to the zoo can fail to be impressed by the size of the feet, especially if two polar bears are playfully patting each other with swipes which would knock a man flying. The soles are well furred, giving a good grip on icy surfaces and probably providing insulation.

Polar bears range through the Arctic Ocean and its coasts. They

have been spotted at latitude 88° N. and are sometimes carried down to
Iceland, Labrador and Japan on icefloes. There are five main popu-
lations which probably remain discrete with only a few individuals
passing between them. These are: (1) Svalbard, Franz Josef Land, East
Greenland; (2) Hudson Bay; (3) Canadian Arctic; (4) Canada–East
Alaska; and (5) West Alaska and East USSR (Cowan, 1972).

Polar bears have always been regarded as a prize trophy. Eskimos
make use of the fur, fat and flesh, but abandon the liver which is so rich
in Vitamin A as to be poisonous. The fur has also been in great demand
among Europeans and, since A.D. 880, when two cubs were brought to
Norway, there has been a trade in live animals for menageries. The
Arctic whalers of the eighteenth and nineteenth centuries looked on
polar bears as a lucrative sideline, alive or dead. The usual method of
capture was to lasso the bear while it was swimming. The result was a
decline in the number of bears which was already noticeable in the
mid-nineteenth century and which accelerated vastly in the mid-
twentieth century when polar bears became targets for sport and could
be shot from aircraft. The polar bear is now considered an endangered
species. The world population is in the order of 10,000. An international research programme was instituted in 1963 to investigate
aspects of polar bear biology which will aid in its conservation, and
international agreement has been reached on its protection since 1976.

A maritime life has, of necessity, imposed on the polar bear habits
unlike those of its land-living relations. It cannot establish a home
range and leads a solitary wandering life. Although a strong swimmer
the polar bear is mainly tied to the pack ice. It prefers less dense ice
where the floes are loosely aggregated and are free to drift rather than
the heavy, compacted polar pack ice (Lentfer, 1972), because seals, the
bear's main prey, are more abundant on loose ice. Where bears are
found on polar pack ice and on fast ice—ice which is frozen in an
unbroken flat sheet near land—they are mainly in transit between
more favourable locations. The circulation of water masses in the
Arctic basin carries the bears with it, occasionally carrying them well
outside their normal range. The result is that bears are more or less
numerous on particular coasts according to whether or not wind and
current have brought in pack ice. As well as this passive movement,
there is an active migration of polar bears. The most famous migration
route is that which leads bears through the town of Churchill, on the
western shore of Hudson Bay.

Individual bears cover great distances. On land or ice, they travel
with an ungainly, shambling walk which is deceptively fast and has

been described as 'distance-devouring'. If in a hurry, the bear breaks into a gallop and attains speeds of up to around 30 kph. Afloat, a polar bear can swim at 10 kph. Submerged except for its head, it forges ahead with powerful strokes of its forelimbs. The hindlimbs trail and act as rudders. The same modifications of the shoulders which enable a brown bear to climb trees are used for hauling a polar bear through the water, and for climbing cliffs and hummocks of ice. It can stay underwater for up to two minutes and while submerged the nostrils are closed and the ears are turned back. The fur is a good insulator in air but poor in water, when heat is lost from the body 25–50 times faster than on land. When afloat the polar bear relies on the thick layer of subcutaneous fat for warmth (Scholander *et al.*, 1950) and, on emerging from the water, shakes itself like a dog.

As a family, the bears are carnivores which have adopted a mainly vegetarian way of life, reflected by changes in their teeth. In entering a polar habitat, the polar bear has, of necessity, changed back to flesh eating and its teeth have reverted to the carnivore pattern of longer canines and taller cusps on the molars. The main food is seals but almost anything edible is acceptable when seals are hard to obtain. Washed-up whale carcasses will attract a crowd of bears, hunters' traps are robbed of their catches, fish are scooped out of the water and birds' nests are robbed. There are a few reports of muskoxen and caribou being killed and eaten and young walruses fall victim if they are not protected by the adults. Lemmings are caught when abundant and bilberries and grass are eaten in autumn. A serious problem for Arctic travellers is the bear's habit of tearing open food stores and devouring the contents. The bear is quite unselective and a list of bizarre items recovered from polar bear stomachs includes coffee, tobacco, sailcloth, adhesive tape, engine oil and an American flag.

When hunting for seals, the polar bear seeks its quarry by scent and captures it by stealth. There is an Eskimo story that a bear will cover its black nose with a paw while stalking a seal. Alwin Pedersen (1966) has described a polar bear's attack on a seal. The bear was on a floe about 300 metres from the shore, where the seal was basking. 'Approaching the water's edge backward, he gripped the ice with his front paws, and then put first one hind-leg then the other into the water. Finally, he let go his grip on the ice, and let himself sink until only his muzzle remained above the surface. This seemed a very clumsy method, but he succeeded in this way in getting into the sea with least noise ... When he was ten yards from the ice floe he lifted his head cautiously, just enough for me to see his ears, and then dived. Suddenly the thin skin of ice which had formed round the floe shattered into pieces, and the head of the bear appeared exactly underneath the seal, which was completely taken by surprise. Before it could make the slightest movement the bear brought his paw on to its head and killed it with a single stroke.'

At other times, the bear waits patiently by a seal's breathing hole, Eskimo-style. As soon as the seal pops up, it is seized and dragged out. Belugas, or white whales, are caught in the same way when they are trapped in an opening in the ice. The whale is killed by a blow to the head as it surfaces and is dragged out of the water even though it is many times heavier than the bear (Kleinenberg *et al.*, 1964). As much as 20 kilograms of food are eaten at a sitting and if hunting is good, the bear eats the blubber and abandons the carcase to Arctic foxes, ravens and gulls which have been following it in expectation of just such an opportunity.

In spring the bears turn to their most abundant source of food. Ringed seals bear their pups on fast ice. Each female seal excavates a den under the snow which connects to the sea by a hole through the ice. Her pup stays here till it can swim, in apparent safety, but polar bears smell out the pups through a metre of snow. The roof of the den is torn open and the defenceless pup is lifted out.

At the end of the summer most polar bears are on dry land because the pack ice is largely dispersed or melted. The bears then seek crops of crowberries, cranberries and bilberries, on which to gorge themselves. Grasses, sedges, mosses, and lichens are also eaten at this time and a bear will dive for seaweeds growing in shallow water.

An orgy of feeding at the end of the summer is needed for the manufacture of reserves of fat. Food will not be easy to find during the winter and the bears conserve their resources by lying up in dens. Male bears are active throughout the year but pregnant females den up for the whole winter. About October the bears search for a deep snow-drift, preferably with a southerly aspect, and burrow into it. The den may be among hummocks of ice but more often the bears find a site along the shore or, rarely, far inland. (On Svalbard they climb 600 metres to find the first autumnal snow drifts.) The den is a simple chamber about 3 metres across and 1·5 metres high, connected to the outside by a passage 1–2 metres long. A pile of snow at the entrance discloses the location of the den. The snow walls give excellent insulation and, if the bear digs down to bare earth or rock, it will construct another chamber leading off the original so that it does not have to lie against the bare, cold rock (Kistchinski and Uspenski, 1972). The temperature inside may be 21° C higher than the outside air temperature (Harrington, 1968). In Manitoba, polar bears dig dens in peat and sand so they can avoid the summer heat. These dens are used as the basis of winter dens but a new chamber is built in the overlying snow (Jonkel *et al.*, 1972). The den usually has a ventilation hole and sometimes the entire roof may be eroded by storms so that the bear is exposed.

Mating, details of which are not known, takes place in March or April, and implantation is delayed for about twenty-eight weeks. The female bear gives birth in the den in late November to early December (Harington, 1965). The usual litter is of two or three cubs; there are rarely four and young mothers (in their fifth year) may have a single cub. They are the size of guinea pigs at birth, weighing less than 0·75 kilogram, and are blind and deaf. The ears open at four weeks and the eyes at five weeks. By six or seven weeks the cubs are beginning to crawl. The mother stays with them in the den until March or April when she forces her way out to break her long fast on grass, moss and berries which she finds under the snow. Departure from the den is stimulated by favourable environmental conditions rather than by the cubs attaining a particular stage in development. The family stays at the den or in a nearby temporary den until the female leads her cubs down to the sea, across the fast ice and on to loose pack ice. This can be a long journey so the family rests up at intervals in shallow holes dug by the female and she sometimes gives the cubs a ride when they are swimming together. The family stays together over a year and only during their second winter do the cubs learn to catch seals.

All large carnivores are a potential threat to human safety and the polar bear is no exception. There is little chance of escape on the flat expanses of ice and snow from an animal which has a superlative nose and can outrun the fastest man. Cases of man killing are rare but the risk is always present because the behaviour of polar bears is unpredictable. The bears are curious and having once seen or scented a man, they approach to investigate him more closely, which is when the risk of attack is greatest.

On the journey across the Arctic Ocean, Wally Herbert's expedition frequently met polar bears which 'just ambled towards you with a completely fearless expression on their faces ... [They] don't seem particularly interested in you at all, but they keep heading in your direction' (Herbert, 1969). Attempts to drive the bears away by shouting failed and they had to be shot if they came within about 6 metres—the bear's curiosity may run to testing the edibility of human flesh. In the same latitudes, one of Nansen's men was bitten in the side by a polar bear but he drove it away by smashing a lantern on its head (Nansen, 1898).

Polar bears have been known to follow a sledge party for several kilometres but sometimes a bear may turn and flee at the first inkling of human presence. Rutting males, females with cubs and bears wounded or very hungry are the individuals which are usually said to be most likely to be dangerous, but Alwin Pedersen's experience (1966), is that females and wounded bears do not attack unless cornered.

The Raccoon Family
(Procyonidae)

The procyonids—there is no popular name—are not familiar to most Europeans. They are small carnivores living mainly in the forests of the New World. The family is represented in the Old World by the lesser or red panda of the Orient and, if it is to be included in the family, the giant panda of China. Excluding the giant panda (now considered to be more closely related to bears), the procyonids are a fairly homogeneous family. The most familiar procyonid is the raccoon of North America.

Procyonids are short-legged and plantigrade with five toes per foot and the carnassial teeth are not well developed. The tail is long and often marked with rings. Many species are arboreal. The diet is omnivorous, consisting of vegetable food and small animals and this is reflected in the flattened teeth. Of the sixteen species, the only European representative is the raccoon which has been introduced from North America.

RACCOON
(Procyon lotor)

The raccoon was introduced from North America to Europe as a fur bearer. One suggestion is that we have to thank Hermann Goering for its presence. Apparently he brought raccoons from America to farm them for their fur but the project failed and eventually two pairs were released into the wild (Anon., 1974).

The raccoon is an appealing animal best known for the bushy, ringed tail, which is left trailing on 'coon-skin' hats, and the black 'robber's mask' over the eyes. The fur is generally long and grey to black giving an overall dirty coloration. The head is fox-like with pointed ears and sharp muzzle, and there is a tuft of hair on the cheeks. The feet have long toes and those of the forefeet are very dextrous. The head and body length is 48–70 centimetres, tail length is 20–26 centimetres. Weight is about 15 kilograms.

The native range extends from Canada to Panama but in Europe introduced raccoons have spread from the Eifel district of Western Germany, along the Mosel Valley and they have turned up in the Netherlands and Luxembourg. They have also colonized Byelo-Russia. The original habitat in North America was wooded or brushy country but where the ground has been cleared, the raccoon has learned to live in open country. This adaptability is probably the secret of its success in colonizing Europe. In both continents, the raccoon successfully lives in close quarters to human beings and is often a scavenger around farms, campsites and even in towns.

Raccoons are solitary, except when a family is living together, each occupying a home range of about 40 or more hectares, with a den in a hollow tree, rock crevice or old badger and fox holes. The home ranges of individuals overlap considerably (Stuewer, 1943). Raccoons are largely nocturnal and activity is reduced in winter, although there is no hibernation. They sleep for longer periods in snowy weather but towards spring they are forced into activity as the autumn's accumulation of fat is used up (Novikov, 1956).

Mating takes place in February or March, after the winter dens have been evacuated. Ovulation is induced by copulation, as in marten and mink, and the cubs, on average four to a litter, maximum eight, are born after a gestation of 63–65 days. They have a coat of fuzzy hair and already possess the typical black mask. The eyes open at two or three weeks, the first solid food is taken at seven weeks and weaning is accomplished at sixteen weeks (Ewer, 1973). The female suckles by sitting back, in the manner of a bear. The first journey from the nursery den is made between six and nine weeks and the family stays together until the end of its first winter.

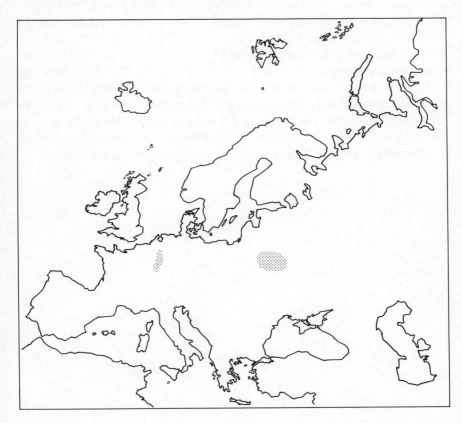

Although primarily a carnivore, the raccoon takes a very wide range of foods. It is less arboreal and more terrestrial or aquatic than other procyonids and this is reflected in its diet. Crayfish and molluscs are favourites but any other small animals are taken, including frogs, earthworms, insects and small mammals. The nests of birds, both on the ground and in the trees, are robbed and raccoons can be a nuisance on poultry farms, as they also climb trees to take roosting chickens (Novikov, 1956). They also plunder crops as vegetable food, mainly fruits, nuts and seeds, are commonly eaten. According to Novikov, Russian raccoons eat as much vegetable as animal food, except in spring when fruits are not available and small mammals are eaten exclusively.

The aquatic habits of the raccoon have led to its most famous quirk of behaviour. It is renowned for washing its food before consuming it, a trait which has resulted in its scientific name of *Procyon lotor* and in the common names *Waschbar* in German and *raton laveur* in French. Although it is sometimes said that raccoons always wash their food, this habit has not been recorded in the wild and Lyall–Watson (1963) showed that the food is not washed but immersed, manipulated and retrieved. He suggests that the behaviour should be called dowsing and that it is abnormal behaviour caused by the captive animal being given its food, literally, on a plate. It does not have to forage, so it goes through the motions of searching for food in water, where much of its natural food comes from. This is reminiscent of captive cats tossing dead animals and pouncing on them as if pretending to hunt.

The Weasel Family
(Mustelidae)

The Mustelidae is translated variously as the weasel, stoat or marten family. It depends on which is the most familiar animal, but all three are 'basic mustelids' with long, slender bodies, short legs and they are specialized hunters. From this basic pattern, the mustelids have evolved a variety of specializations. The otters and mink, still short-bodied and slender, have become aquatic. The pine marten is arboreal, the South American tayra is a fruit eater and the badgers are heavy-bodied omnivores which are good diggers.

Except for the badgers and the tayra, the mustelids depend almost entirely on animal food and, as a family, they are more carnivorous than the dogs. The jaws are short with powerful muscles and are armed with sharp teeth. Prey is typically killed with a well-aimed bite to the back of the neck. Some species can tackle prey larger than themselves. The feet are plantigrade or digitigrade, with five toes registering in footprints. The fur is soft and dense, and some mustelids, notably mink, martens and the sable, are greatly prized as furbearers. Except for the badgers, mustelids are solitary. Secretions from the anal glands are used to mark territories and, as in skunks, as a weapon of offence. The litter is usually born after a period of delayed implantation.

There are sixty-seven species of Mustelidae arranged in five subfamilies. The Mustelinae includes the weasels, martens, wolverine, tayra, sable and others; the Mellivorinae consists of a single species, the honey badger; the Mellinae includes the badgers; the Mephitinae, the skunks; and the Lutrinae the otters.

Distribution is mainly over the northern hemisphere and there are no native mustelids in Australasia. There are eleven species native to Europe plus the introduced American mink. The sable has become extinct in Europe.

WEASEL
(Mustela nivalis)

The smallest of European carnivores, the weasel has a reputation for ferocity far out of proportion to its size. It is said to be able to pass through a wedding ring and, as its body can squeeze through any opening that its head has already negotiated, this seems likely. The photo (pp. 112–3) suggests that the saying is correct but a tame weasel failed, despite strenuous efforts, to get into its sleeping box when the entrance was reduced to wedding ring size. Nonetheless, weasels are small enough to slip down mouse holes and agile enough to raid birds' nests and there can be few places where potential victims are safe from attack.

In overall appearance, the weasel is similar to the stoat. Both are small, lithe-bodied and short-tailed, but the weasel is much the smaller animal. Gilbert White aptly described the weasel as 'the little reddish beast not much bigger than a field-mouse, but much longer'. The head and body of adult males range from 21–23 centimetres with a tail of 6–6·5 centimetres. Measurements of females are 16–19 centimetres and 4–5·5 centimetres respectively. Weight range is 60–130 grams for males and 45–60 grams for females and it is important to note that the mean weight of females is half that of males. The fur is reddish-brown above, white underneath, and the weasel lacks the black tail-tip of the stoat. Moreover, the boundary line of brown and white are nearly always irregular, compared with the straight demarcation on the flanks of the stoat. There are often blotches of colour on the under-parts, notably a spot on each side of the throat, which are absent in the stoat. The pattern of markings enable individual weasels to be identified (Linn, 1962). The coat turns white in winter in the northern parts of the range. White weasels are very rare in Britain.

The systematics of the weasel have been complicated by two problems which are not fully resolved. The European weasel *Mustela nivalis* is usually separated from the American least weasel *M. rixosa* but they may be conspecific (Reichstein, 1957). Within Europe there are frequent reports of a species of dwarf or pygmy weasel, considerably smaller than the normal size. In Britain this has been called the cane weasel, least weasel, kine, mousehunt or mouse weasel. Gamekeepers and others are emphatic that they are not merely describing the smaller female. However, as weasels bear two litters per year, the first maturing rapidly and the second more slowly, there is a considerable size range in weasels and the cane weasel could be no more than immature individuals (Owen, 1965). Van den Brink (1967) describes a pygmy weasel of northern and central Europe very much like the American least weasel. It is about 4 centimetres shorter in overall length, with a regular demarcation between brown back and white belly, usually no throat spots and white feet, but he does not regard it as a separate species.

The weasel ranges across Europe and Asia to Japan, from the Siberian coastline south to Afghanistan. It is also found in North Africa, but it is absent from Ireland where the stoat is, confusingly, called the weasel. The habitat is woodland, farmland, moorland and tundra, wherever there are voles and mice—including towns. Weasels were introduced, along with stoats, to New Zealand in 1885. After an initial abundance, their numbers declined and they are now rare (Hartman, 1964).

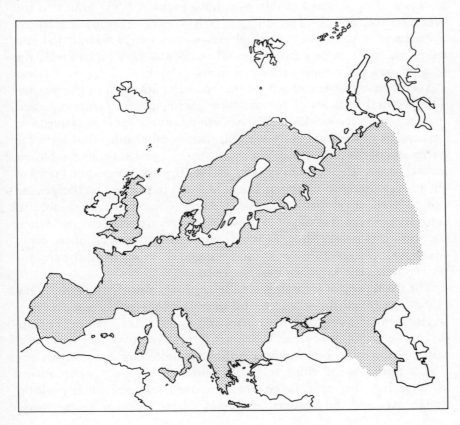

Because of their small size weasels are not often seen, although they are active by day as well as by night. The best time to see them is when they cross roads and paths or climb walls and tree trunks. Their speed of movement means that there is barely time to identify the animal but inquisitiveness may draw a weasel to within a metre or two. It comes gradually closer making the best use of cover so that the onlooker sees no more than a triangular, mouse-like head appearing from behind stones or leaves, unless the ground is open, when the characteristic bounding, hump-backed gallop or gliding, straight-backed scuttle can be seen.

The weasel has been described as bloodthirsty, vicious and murderous; epithets which are anthropomorphic and attribute to the animal a love of killing which cannot be substantiated objectively. Nevertheless, these words convey the impression of a hunter whose persistence and tenacity outweigh its diminutive size when procuring prey much larger than itself. Only by dogged and repeated attacks can a weasel bring down a rabbit, weighing perhaps 1,500 grams to the weasel's 100 grams. The dogged trait is well exemplified by an eyewitness account in a newspaper of a weasel which was carried into the air on the back of a gull. Even when the gull dived into a lake, the weasel's jaws were not dislodged from its neck.

Such behaviour is assisted by the muscular strength of the weasel. The pectoral muscles of the shoulder girdle and the temporalis and masseter muscles working the jaws are relatively more massive, as a percentage of the total body weight, than in other mustelids with the exception of the powerfully built wolverine. These muscles enable the weasel to wrestle with heavy prey and it does not dislocate its jaws in the process because bony flanges hold them in place. The flanges are not so well developed as in the badger, where it is impossible to separate jaws from skull, but the weasel's jaws are securely held when the mouth is nearly shut, as when fastened to the neck of its prey. Moreover, the jaw muscles are so arranged that they exert maximum power when the jaws are nearly closed.

The main prey of weasels is small mammals, principally voles. Day (1968) examined the stomach and intestine contents of weasels collected from all over the British Isles and found that 57 per cent of the diet was composed of small rodents. Of this sum, over half were field voles; wood mice, rats and house mice accounted for the rest. Walker (1972) found wood mice accounted for half the prey items in his sample, but this was taken from weasels trapped mainly along hedgerows and the borders of woods where wood mice are more

A marginal member of the fauna of Europe and most carnivorous of the bear family, the polar bear has a marine life and swims powerfully in the cold Arctic seas.

Forced back into mountain retreats, the brown bear is too large to live comfortably alongside man. Although not much of a hunter, it does sufficient damage to bring down retribution.

The red fox is the most adaptable of carnivores and has taken to living in towns where garbage provides it with plenty of food.

The weasel is the smallest European carnivore and is a specialist hunter which can overcome prey far larger than itself.

A stoat in its winter coat, called ermine. The precise function, whether for camouflage or to help keep warm, is not certain.

A polecat standing on its hindlegs and craning for a good view, in the typical attitude of short-legged carnivores.

The beech marten prefers open country and a liking for rocky places earns the alternative name of stone marten.

The bushy tail of the pine marten acts as a balancer when the pine marten climbs trees and jumps from bough to bough.

The boldly contrasting black and white stripes on a badger's face were once thought to act as camouflage in moonlight. It is more likely that they are designed to make the badger conspicuous.

Well adapted for an aquatic life, the otter has waterproof fur, webbed feet and a stout tail for propulsion.

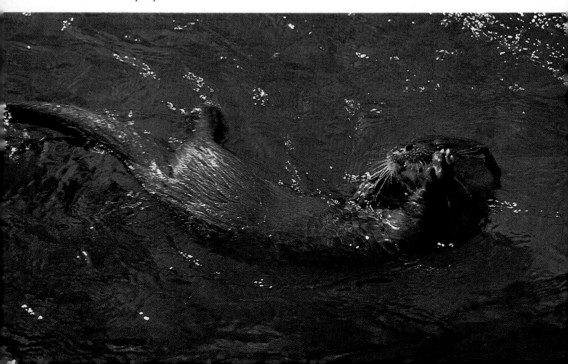

The lynx is a handsome cat with 'sideburns' and tufted ears.

Like a delicate tabby cat, the feline genet is at home in trees. It is normally abroad at night.

The strange-looking raccoon dog has been introduced to Europe from the Far East, for the sake of its fur.

abundant. In addition the sample was collected in autumn and winter when lack of ground cover would make the mice more vulnerable. In a rather similar habitat, in the north of England, few wood mice were eaten and bank voles completely ignored, despite their abundance, while in an Oxfordshire woodland bank voles are the most important prey (Moors, 1975).

Of larger rodents, rats are taken in numbers only when the weasels are frequenting farmyards where they are abundant. Squirrels have the advantage in arboreal encounters, although Parovshchikov (1963) gives two instances of weasels eating red squirrels. Considering their small size, weasels kill a surprising number of rabbits and hares, although the larger male weasels account for twice as many as the females. These lagomorphs comprised 19 per cent of Day's sample. Moles are pursued underground and shrews are caught in small numbers but often not eaten. Captive weasels will eat freshly killed shrews if they are very hungry. Outside the British Isles, mammalian prey includes lemmings, Brandt's vole, common vole, root vole, ruddy vole, garden dormouse and striped field mouse (Parovshchikov, 1963). Records of elk in weasel stomachs must refer to carrion!

Birds, mainly passerines or perching birds, are an important food (15 per cent in Day's sample). Presumably they are stalked on the ground or in the bushes. Parovshchikov records goldcrests being killed when a snowstorm forced the birds to roost near the ground. Nests are vulnerable and weasels are small enough to raid hole-nesting birds such as tits.

A particular study has been made of Wytham Wood, Oxford, where it was discovered that some weasels learned to associate nest boxes with food (in the experimental study on tits there). The tit broods seem to attract weasels most when they are hungry and calling noisily (Perrins, 1965). In some years 70 per cent of the clutches in these nest boxes are lost to weasels (Krebs, 1970). Predation by weasels became acute after the wood was cleared of rabbits by myxomatosis in 1957. Predators previously relying on rabbits turned to catching rodents and weasels were forced by the new competition to hunt elsewhere. Now weasels are the major cause of nesting failure in the wood and their predation is most severe when nesting density is high (Dunn, 1977).

Pheasant eggs are stolen (Day, 1968) and two golfers once came across a weasel wandering about with its head stuck in a pheasant egg. It was freed with a well-placed kick (Pentecost, 1963). Cold-blooded prey includes small numbers of insects, such as beetles and earwigs, and a few worms. Parovshchikov found that frogs, common lizards and sand lizards were taken.

The weasel launches itself at the victim's neck and sinks its long, sharp canines deep into the flesh. A vole or mouse is killed instantly as the teeth penetrate the skull and destroy vital brain tissues. With larger prey, repeated attacks are needed and the weasel must hang on until the victim succumbs. If possible the weasel tries to throw its victim off its feet and then rakes it with its claws.

It used to be said that weasels sucked the blood of their victims but this is an incorrect deduction from observation of the weasel's hunting behaviour. The eyewitness sees the weasel holding fast to the victim's neck, from which blood is flowing copiously and covering the weasel's snout. When the weasel is disturbed and retreats an examination reveals twin punctures caused by its fangs and the inference is that the weasel has the habits of the legendary vampire. In fact it is about to commence its meal of flesh, first having dragged its prey into the seclusion of the undergrowth. It may be as well to lay to rest another falsehood about weasels. They are not cruel. It seems the victim is in a state of shock and feels no pain, which was the experience of David Livingstone when he was mauled by a lion.

A hungry weasel will consume almost an entire vole or mouse in one sitting, leaving only the front part of the skull, the stomach, feet and tail (Linn, 1962). The daily requirement is 25 per cent of the body weight or about two voles. Stoats eat a similarly large amount of food, which is two or more times more than that eaten by other carnivores. This could be because of the shape of their bodies (p. 91). If food is plentiful, prey is cached around the nest for later use. Parovshchikov found a cache containing sixty-nine voles, another had forty-seven frogs and a third had thirty-four sand lizards.

Weasels are more abundant than the similar but larger stoats. Each individual weasel needs only one-eighth of the hunting range required by a stoat although it is only half the weight. Lockie (1966) suggests that this is due to the weasel's smaller size giving it access to a far larger supply of prey. It can penetrate the runs of mice and voles and flush them from their nests, whereas the stoat is forced to hunt above ground. Moles, too, are vulnerable in their underground retreats and weasels are probably their main predator. The same division of hunting behaviour occurs when the ground is snow-covered. Nyholm (1959) has watched stoats hunting over the snow. They plunge through the crust when the vole has been located or wait for one to emerge. By contrast, the weasel slips down the burrows where voles are feeding on buried herbage. Although weasels have access to food denied to stoats, the dependence on voles makes them vulnerable to the well-known 'crashes' of vole populations.

Each weasel lives in a territory where it will be familiar with the terrain and good hunting places. Within the territory it has a den, sometimes lined with the fur of its victims. The den is used for a short time before being abandoned for a new one. According to Lockie's study of weasels and stoats in a Scottish forestry plantation, the weasel's area of activity should be considered as a territory rather than a home range because its boundaries do not overlap those of other weasels, but in Wytham Wood, Oxford, weasels had overlapping boundaries (King, 1975b). The territories are defended by marking with scent from the anal glands and by overt defence but mainly by mutual avoidance. A chance, but enthralling, sight is to see two embattled weasels rolling over and over in a ball, locked together until one breaks away and retires, squeaking, or to see the territory owner chasing an intruder pell-mell over the ground and out of sight.

The territory is probably not seriously defended because of its large size, so transients can pass through without much trouble but the owner can prevent another weasel settling. If a territory-holding weasel should die its place is taken by another who takes up residence in almost the same area. If no newcomer appears, the neighbours gradually extend their own boundaries until the vacant territory has been completely swallowed up. On one occasion, several of Lockie's weasels went missing or died. Their absence left such a large vacuum that it could not be filled and the territorial system broke down. The weasels extended their territories and, meeting no opposition, kept moving. A breakdown of territorial system when the population is sparse has been found in other mustelids. It would be interesting to know what effect this has on breeding.

The newcomers who take over vacated territories come from a reserve of mobile animals. They wander across country, chivvied by territory owners whenever they try to settle and, with no chance to become familiar with the terrain, they must have difficulty hunting. Often the transients are sexually immature, with testes remaining small during the breeding season, and maturity comes with the acquisition of a territory. Transient weasels are not necessarily young animals. A weasel can spend its life as a landless wanderer as, to obtain a territory, it requires the luck of being on the spot when one becomes vacant or it has to displace an incumbent.

The territories of Lockie's male weasels ranged in size between 1 and 5 hectares. The females occupy smaller territories within those of the males. In an English woodland territory size was 7–15 hectares for males and 1–4 hectares for females (King, 1975b). One male had a lifetime range of 26 hectares but only used 12–13 hectares at a time. For much of the year the female is subordinate to the male and she has to defend her territory against him.

Aggression dies down during oestrus to allow mating. Hartman's (1964) captive male weasels trilled with pleasure when a female was introduced to their cages, and copulation lasted for two or three hours. As pregnancy proceeds, the relationship changes and the female assumes a dominance over the male which lasts until her young become independent. Relations between the sexes are consequently rather strained and if food becomes scarce or the population very dense, the females are driven away by the males. This looks like a very odd form of population control and the fate of the dispossessed females is unknown.

Weasels are unusual among carnivores in having two litters per year, at least in the mild climate of southern England (King, 1975b). There is no delayed implantation. Pregnancy lasts five to six weeks and lactation ten weeks (Deansley, 1944 and Hartman, 1964). Females come into breeding condition in February or March, so the first litter may be born in April or May but some first litters are not born until August. Provided that food is plentiful, weasels which have an early first litter bear a second litter in July or August. Kittens of the first litter grow rapidly and may become sexually mature and breed at the end of the summer. Second litters grow slowly and mature in the second year.

The litter of four to six kittens is born in a nest, lined with dry leaves, mosses and fur from prey. It is located in a hole in a bank or in a hollow tree and is hard to distinguish from a mouse hole. Development of the kittens is only known from captive females (East and Lockie, 1964, Hartman, 1964 and Lingard, 1965). The kittens are no more than 4·5 grams when born. The pink skin is barely covered with short grey hairs, but by the end of a fortnight the kittens are grey above, white underneath and the grey gradually turns brown by three weeks. On the seventeenth day the ears open and the milk canines erupt. A day or two later sees the kittens crawling in straight lines, instead of in aimless circles, and five days later they are becoming active. The eyes open at four weeks but a few days earlier the mother brings solid food to supplement her milk and the kittens become house-trained by leaving the nest to defaecate. When fleas become too numerous the mother moves her family to a new nest.

At five to six weeks the permanent canines erupt and only a few days later captive weasel kittens can kill mice. One kitten was observed to hesitate on the first confrontation but, once it had seized the mouse by the head, it quickly despatched it. A fair amount of practice is, however, needed before hunting and killing become proficient. Wild weasels would be hunting in family parties at this stage. With their quick, darting movements and sudden appearances from behind stones and logs, a female weasel with five or six kittens looks like a swarm of dozens and several families may combine to make a formidable pack. The family breaks up at fifteen weeks through increased aggression among its members. One captive litter even killed its mother.

Despite its small size a weasel is a formidable enemy when cornered. Hissing, shrieking and rearing up like an enraged snake, an alert weasel is a challenge even to a bold predator. The inclusion of weasels in lists of prey of birds of prey and carnivorous mammals shows that the weasel must sometimes be caught off guard. While it is easy to visualize a buzzard or an owl dropping, unsuspected, on a weasel hunting with its nose to the ground, an attack by a stoat or kestrel would seem to be too even a match, unless the weasel were diseased or lame.

STOAT
(Mustela erminea)

Much of what has been written about the weasel is applicable to the stoat. In many respects the habits of the two are similar except in respect to the stoat's greater size. For instance, it is less able to penetrate rodent burrows. Male stoats have a head and body of 22–29 centimetres and a tail of 8–12 centimetres. Weight range is 125–300 grams. Females are smaller and approach the size of a male weasel. Size apart, the immediate identifying feature of a stoat is the black tipped tail. Closer examination shows that the boundary of the red-brown upperparts and the white underside is even rather than irregular as in the weasel. There are no blotches on the underside but there is often a little white on the face.

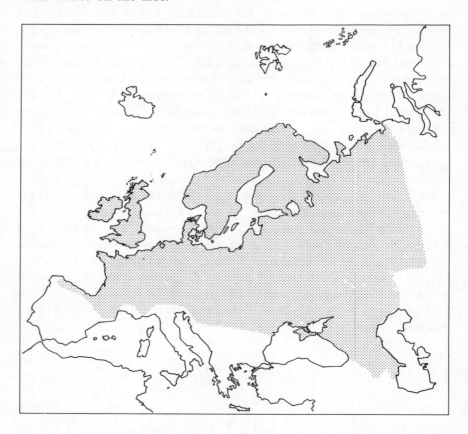

The stoat's range overlaps that of the weasel but, in Europe, it is found in Ireland, where it is called the weasel, but is missing from the Iberian peninsula, south-east France, Italy and much of the Balkans. It is also found in North Africa and ranges across Greenland and North America where it is called the short-tailed weasel.

In the northern parts of its range, the stoat turns white in winter except for the black tip of its tail. In this condition it is known as ermine. According to legend the stoat is so jealous of its white coat that it will die rather than sully it. Ermine pelts have been valued for centuries and are used for trimming ceremonial robes of the nobility. There is a well-known picture of Queen Elizabeth I of England with an ermine on her sleeve as a tribute to her spotless virginity. The artist has flecked the ermine's coat with black marks without realizing that the black flacks on ermine trimming are the tails of stoats.

The exact distribution of the ermine form of stoats is not known. In Britain, for instance, stoats living in Scotland and northern England turn white but those in the south of England do so only rarely. Colour change is extremely rapid, being accomplished within three weeks or as little as three days. It used to be thought that this entailed an actual bleaching of the hair in autumn but the speed of change is due to a new coat forming under the old one, which is then moulted *en bloc*. Moulting takes place twice a year, in autumn and spring. In Denmark, stoats start to turn white in the second week of November (Fog, 1969), but some are still in the full summer coat in mid-December. The spring moult is more prolonged. It starts in mid-February, the first full summer coat is seen in April and white animals are still in evidence in mid-March.

White fur is a characteristic of many Arctic or north temperate mammals and birds. The polar bear is permanently white, while others, including Arctic fox, Arctic lemming, mountain hare, ptarmigan, and the stoat and weasel, vary between white and shades of brown. There are two possible explanations. The most persistent explanation is that the white is a camouflage to conceal the animal against a background of snow and ice but there has also been a suggestion that, in seal pups at least, a white coat is a means of keeping warm. The theory is that the white hairs act like the glass of a greenhouse. Sunlight can penetrate to warm the body but the hairs prevent heat in the form of long-wave radiation from escaping. More recent experiments seem to discredit this theory. The difficulty of applying the camouflage idea to stoats is that they turn white in places where there is little snow. White stoats on the mild west coast of Scotland, where snow cover is rare, are the reverse of camouflaged. A white stoat stands out clearly against a snow-free background but this may be no great disadvantage to an animal which hunts through dense herbage and down dark burrows. It could be that colour change is under genetic control and that these stoats have not adapted yet to milder climates but this is unlikely to be the entire explanation because patterns of colour change can be altered experimentally.

The occurrence and timing of colour change involves two physiological factors: the presence or absence of pigment in the new hair and the replacement of the old hair with the new at the moult. Both are under the control of the pituitary gland and are influenced in turn by environmental factors: change in day length and temperature, which act on pigment and moult in a rather complicated fashion. Cold weather alone turns English stoats white in winter but prolonged cold does not prevent them returning to brown in spring. American white stoats were kept under artificially lengthened daylight. Some were maintained at 70° F; others at 20° F. The longer days made them moult to brown earlier than normal in spring but those kept at a lower temperature required longer for the process to start (Rust, 1962). A colour change at the autumn moult seems to depend on whether the necessary requirements of temperature and day length for pigment inhibition in the new hair have been fulfilled when moult takes place. There are also genetic factors at work. English stoats do not usually change colour but an alpine stoat from Switzerland continued to do so even when kept in mild English winters (Rothschild, 1957).

In the matter of diet the stoat is much the same as the weasel but it takes fewer voles and more larger animals such as rabbits and hares, and pigeons and gamebirds. Stoats even eat weasels, and a macabre item in several lists of the various foods found in the stomachs of stoats is claws from their own legs which they have gnawed off when caught in a trap. In the same study that revealed the diet of British weasels (p. 80), Day examined an equal number of stoats and found that small rodents occurred in the stomachs and intestines of half the weasels but only one-fifth of the stoats. On the other hand the proportions were almost reversed for rabbits, hares and birds. Half the small rodents were field voles but the stoats had eaten more bank voles and wood mice. Surprisingly, the smaller females ate much the same as the males but they are probably not so good at catching rabbits. Rabbits are attacked not only in the open but are also chased down their burrows and there is a record of stoats preying on nesting sand martins (Mead and Pepler, 1975).

Differences in diet are the key to the co-existence of such similar animals as the stoat and weasel. The weasel's emphasis on chasing mice and voles down their holes and the stoat's preference for larger prey lessen competition, although there may have been a confrontation when myxomatosis forced stoats to join the band of rabbit hunters which had to turn to rodents. There must also be some competition between female stoats and male weasels and more subtle differences in diet may not be revealed in a gross study of stomach contents. For instance, it seems that weasels raid nest boxes in Britain whereas stoats raid them in Germany.

Stoats, and weasels, are considered to find their prey mainly by smell. Considering their low stature and their habit of searching for food in undergrowth and rank herbage or down burrows, it is difficult to see how vision could be of much use except for guiding the final pounce. The search is aimed at finding a scent trail which can be followed unerringly to its source. There is a story of someone who watched a stoat single out a rabbit from a group feeding in a quiet country lane. The rabbit, with the stoat following about 5 metres behind, described three circuits of 20 metres up and down the lane. At some points the rabbit and the stoat ran past each other, no great distance apart, yet the stoat made no attempt to cut corners. It was intent on following the scent trail.

A stranger description of hunting behaviour, which is also told about weasels and foxes (p. 36), is that of their charming their prey. There are so many eyewitness accounts that it may well be a fairly common trait of these animals. There are, after all, not that many eyewitness accounts of small carnivores hunting by conventional tracking or chasing. One particularly vivid description tells of a stoat squirming on the ground, somersaulting and throwing itself backwards in the air. Gathered on the nearby hedge and the ground underneath were a dozen or so greenfinches. They seemed to be fascinated by the stoat and some even appeared quite dizzy. One was described as looking 'positively intoxicated' with its wings spread and drooping. Feeling sympathy with the birds, the eyewitness ran forwards shouting but the 'intoxicated' greenfinch could not move for a second or two.

If it had not been for the timely interference, one of the greenfinches would have fallen an easy prey to the stoat. This account is more detailed than most and it shows the general pattern. What would be interesting to know is whether these accounts describe a stoat setting out deliberately to lure its prey by gymnastic antics or whether they are incorrect interpretations of a chance meeting. Stoats are very playful animals and will romp together or singly. That birds are curious is shown by an incident with two grey squirrels kept in a cage at the bottom of a garden.

Squirrels play like stoats, on the ground, indulging in almost boneless contortions. The garden was the home of a variety of ducks and chickens and, one day when the squirrels were playing, they unwittingly attracted the attention of the poultry who lined up along the front of the cage, beaks to the wire, and watched for ten minutes. This sort of incident happened many times and the pattern was for one bird to be attracted if it came within 5 metres of the squirrels' cage. Its interest in the squirrels caught the attention of the other birds who would wander in from a radius of 50 metres. So here is the basis of the stoat's charming: a playful animal attracting curious animals. There is no reason why the stoat should not learn, after accidentally attracting prey in this manner, that cavorting in front of an audience brings an easy meal.

As far as is known, the social life of stoats is essentially similar to that of weasels although there are important differences in breeding. As individual stoats require a larger range than weasels, they are widely scattered and are less abundant. One habit, which will evince the same incredulity from hard-headed people as the stoat's charming, is the story of the stoat's funeral. There are fewer accounts of funerals than of charming but they are described in detail and keep on cropping up. One such was described in *The Irish Times* for 13 February 1952. A car driver noticed what he thought was a monster snake crossing the road but, on drawing near, realized that it was a procession of about a hundred stoats, an incredible number in itself. The leading four were carrying the body of a dead stoat. The driver attempted to follow but turned back when the stoats hissed at him.

Animals as diverse as squirrels and rats, elephants and monkeys, have been seen carrying dead companions but the reason is a complete mystery, unless the animals are unable to tell that their companions are dead and are endeavouring to assist them in the way that elephants and dolphins assist wounded congeners. Unfortunately this sort of behaviour cannot be subjected to systematic observation or experiment (and maybe rightly so as life would be poorer without mysteries). The figure of a hundred stoats can be dismissed as an exaggeration (a family party or two families combined on the move gives the impression of a swarm) without necessarily destroying the central fact of the corpse being carried. If an anonymous, and third-hand account, in an Irish paper sounds suspicious, an eminent, but understandably also anonymous, authority on British small mammals, has described watching two stoats retrieve the body of a third and drag it into a burrow in the bank. The actions of the stoats suggested very strongly the action of a female stoat retrieving her young, and this may be the root of the so-called funeral—no more than a displaced maternal instinct.

The breeding of stoats differs from that of weasels in the production of a single litter a year and in the occurrence of delayed implantation. There are two peaks of mating, most conceptions taking place either in March or in June and July. Further matings may take place after July but these are infertile. Implantation is delayed and the blastocysts lie free in the uterus until the following spring, when gestation then takes only three or four weeks. Why there should be a period of delayed implantation is a mystery, because it seems feasible for mating and birth to occur in one year. Occasional winter litters are known when the delay of implantation has been presumably omitted.

Birth takes place in April or May and the litter is of four or five, sometimes more. At first the babies have a sparse covering of white hair and the black tip of the tail appears at twenty days. Eyes open at four weeks. The family stays together after it leaves the nest and they quarter the mother's range together, sometimes rushing along in a mass, looking like a piece of sacking blown by the wind. It is sometimes stated that the stoat is less aggressive than a weasel but there are records of stoats attacking men and even of killing dogs (e.g. Malone, 1965).

Weasels and stoats are inhabitants of temperate and Arctic regions, where they survive winters in the tundra. Brown and Lasiewski (1972) point out that the long, slender body is a disadvantage because it will lose heat faster than a more rotund body shape. Compared with a rat of the same weight, a weasel or stoat has a greater surface area, shorter fur and, when curled up, forms a flat disc rather than a more compact ball. Small mustelids must, in consequence, expend more energy in keeping warm. They are active by day and night and so meet a wide variety of prey but a high energy requirement means greater intraspecific competition for food: individuals of each species are competing for prey with their fellows. Brown and Lasiewski suggest that this competition is reduced by sexual dimorphism, the larger males seeking different prey from the smaller females. Furthermore each sex lives in a territory which is maintained against others of its sex, and so reduces the competition between animals of the same size and prey preferences. However, the female also defends her territory against the male, presumably both in defence of her litter and to ensure that he does not eat their prey.

In defence of this theory, Brown and Lasiewski point out that mustelids without long, thin bodies, the badgers for example, do not have marked sexual dimorphism. However, the other slender mustelids do not wholly support this theory. The otter, the largest species with this body shape, shows little sexual dimorphism and, apparently, no difference in diet and mink which are highly dimorphic also have no difference in diet. Pine marten females are only a little smaller than the males but Yurgenson's data (p. 107) show that there is a significant difference in diet: males alone taking large prey such as hares and gamebirds. On the other hand Yurgenson showed that the female steppe polecat has proportionately heavier musculature of forelimbs and jaws which allow her to take the same prey as the larger male.

Clearly the situation is not simple and such factors as the availability of different kinds of prey and competition with other species must be taken into account. Yurgenson points out that pine martens have a wide range of food. Both sexes eat large numbers of squirrels but the rest of the diet is made up of prey of sizes suited to each sex. The steppe polecat has a rather restricted range of food available: both sexes must feed on steppe-dwelling marmots and susliks. With regard to interspecific competition, female stoats are little larger than male weasels and, like them, can hunt down rodent burrows.

EUROPEAN MINK
(*Mustela Lutreola*)

The European mink is a little-known animal compared with its American cousin. The latter has become a domestic animal, having been brought to Europe to be bred for its pelt. Numerous escapes have resulted in its becoming naturalized in many parts of Europe and it has been blamed for displacing the native species, but the range of the European mink had already been decreasing (Novikov, 1956). American mink are described in pages 94–97. The present range of the European mink includes Finland (north to 64° N), north and west France, parts of Poland, Romania, Bulgaria and Yugoslavia. The marshes along the Danube are a stronghold. Mink have disappeared since the early part of the century from Austria, Switzerland, Czechoslovakia, Italy and Hungary. They are still found in Russia and they have spread eastwards into western Siberia within the last century as far as the rivers Ob and Tobal.

The build is of a large weasel: body length 34–40 centimetres, tail 13–14 centimetres and weight 550–800 grams. The coat is a uniform deep brown, relieved only by white on the lower, and often the upper, lip. Van den Brink (1967) notes that the American mink never has white on the upper lip. To some extent the mink is adapted for an aquatic way of life. The feet are partly webbed and the toes bear bristly

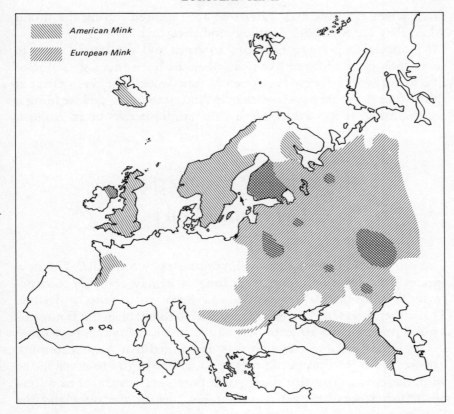

American Mink

European Mink

fur which assist in paddling. The body is streamlined, with dense fur, thick skin and a thick layer of subcutaneous fat (Novikov, 1939). Compared with the otter, the mink is not well adapted for an aquatic life. It lacks the powerful, flattened tail and mink eyes have no adaptations for underwater vision (Sinclair *et al.*, 1974).

The habitat of European mink is small pools and streams or marshes, and is generally being more restricted than the American mink. Otherwise, it must be assumed that the habitats of the two species are very similar. The European mink is said to roam in winter but to establish a home range of 15–20 hectares in summer (Novikov, 1956). Within the range, a den is built among clumps of saplings in natural crevices or in specially enlarged water vole burrows. Activity is concentrated in the hours of darkness but cloudy weather brings mink out because their prey will become active in these conditions. The main prey is small rodents, especially water voles. Fish are taken in quantity but the mink seem to be less dependent on them. Frogs are eaten when available and waterfowl are captured during the moult when they are flightless. Surplus food is cached.

Mating takes place in February to April and the kittens, two to seven, but usually four or five in number, are born after a gestation of five to six weeks. Their eyes open at four weeks, the teeth erupt at thirty-four days, ten days later than in American mink, and weaning is accomplished at ten weeks, so that the family breaks up in August.

AMERICAN MINK
(Mustela vison)

The introduction of an alien carnivore poses two threats. The new species may devastate the native fauna or it may compete too successfully with native carnivores and drive them into a decline. That either threat has become reality is hard to establish but it must be a good principle not to introduce any alien animal which may upset the ecological balance. When it was discovered that American mink had escaped from European fur farms, and had begun to breed in the wild, there was a fear that they would pose a serious threat to waterbirds, fish stocks and native carnivores such as the otter and the European mink. American mink were first brought to Europe in the late 1920s. Wild and escaped (feral) mink were found in Sweden in 1928 and in Britain in 1938, but only later was it confirmed that they were breeding successfully in the wild. The American mink is now one of the most common carnivores in the British Isles and Scandinavia.

The American mink is very similar to the native species in both appearance and habits, but it is rather larger. Chanin's (1976) measurements of American mink in two English rivers average head and body 39 centimetres, tail 19 centimetres and weight 1,080 grams for males; and 34 centimetres, 17 centimetres and 606 grams respectively for females. Unlike the European mink, the American mink has white on the lower lip and chin alone, often with more white between the forelegs. Many colour varieties are bred on mink ranches, and although the true wild form is very dark brown, white, cream and other 'pastel' mink frequently turn up among the feral populations.

The mink has been well studied in North America because of its

economic value as a fur-bearer and is better known in Europe than its
European cousin because of its status as a potential pest. Mink are
largely nocturnal but they can be seen and trapped during the day
(Cuthbert, 1973), perhaps being brought out in bad weather as
suggested for European mink (p. 93). They are restricted to the vicinity of lakes, marshes and watercourses and can occupy small streams
no more than a metre or two wide. Studies by live-trapping in Britain
(Chanin, 1976) and by radio-telemetry in Sweden (Gerell, 1969 and
1970) show that the American mink (and, we may assume, the European mink) is solitary and organizes its life in the same manner as other
solitary mustelids. Established mink live in a defended territory running along the shores of a river or lake. The territory is marked with
spraints like an otter's (p. 133) but the species can be distinguished by
the otter's spraints having a pleasant smell while those of mink are
foul. Mink rarely move far away from the water's edge and males
occupy a length of waterway ranging between 1·6 and 4·4 kilometres
(Chanin, 1976). Females have slightly smaller territories (range
1·2–3·2 kilometres). Not only do the territories of male and female
overlap, there may also be some overlap between neighbouring males.
However, there is a temporal separation and the mink avoid meeting
each other. Within the territory, each mink has several dens in hollow
trees, under exposed roots or in widened burrows of other animals.
Chanin's studies showed that mink occupy a territory for a surprisingly short time (the record was for ten months) and territory
holders may become transients.

Mating takes place in March or April, the antagonism between male
and female being set aside for a while. Ovulation is induced by
copulation with a delay of 35–40 hours. Implantation is also delayed,
but for a variable length of time (Gerell, 1971). The total length of
pregnancy ranges from 39 to 76 days with the post-implantation
period lasting 28–30 days. Average litter size in Gerell's Swedish study
was 3·8, while Chanin found an average of 5·9. Little is known of
family life in the wild but Paul Chanin recorded a wild female mink
leaving her territory while accompanied by kittens. Weaning is
accomplished at eight to ten weeks and the family disperses in August.
Young mink become sexually mature, and may breed, in their first
year.

Despite an obvious correlation between diet and riparian habitat,
the mink is an adaptable, generalist feeder. Chanin has summed up the
mink thus: 'Stoats are better at killing rats and rabbits, weasels at
killing voles, and the otter is a more efficient fish predator than the

mink.' It is also a good tree climber but not sufficiently good to catch squirrels and challenge the pine marten. But the mink has a wider variety of prey available, providing it does not have to compete with other mustelids. Erlinge (1969) found that mink avoided competing with otters for fish in winter by taking up more terrestrial habits. There is less competition in summer when more food is available and the two species live side by side, although otters tend to keep mink out of their favoured habitats (Erlinge, 1972).

The bulk of the diet is small mammals—field and water voles, or rabbits, even shrews may be taken in large numbers—but the precise proportions depend very much on circumstances. In Sweden (Gerell, 1967), amphibians are eaten particularly in spring and autumn when they become a stop-gap between the winter diet of fish and the summer diet of crayfish. Of fish taken, the mink follows the otter's lead of taking whatever is easiest to catch. Some damage is done to stocks of trout but eels are also eaten. Unlike the otter, the mink hunts by plunging in from the bank rather than diving from the surface. It is, therefore, best able to catch those animals visible from the bank. Chanin found that, at a lake fringed with reeds, the mink were eating eels and stickleback which penetrate the reedbeds but not fish of open water because there were no convenient banks to dive from. Avian prey includes rails, ducks, passerines and pigeons. Rails are easier to catch than ducks, and pigeons are caught when they are feeding on the ground. Havoc can be wrought in poultry runs but this is only likely to occur if fences and buildings are inadequate for keeping out predators. Beetles and earthworms are also eaten and even the occasional weasel is devoured (Chanin, 1976).

The results of research into mink diet suggest that this alien is not going to prove so damaging as was once feared. There has not been a wholesale destruction of fish stocks and alarmist predictions of mink completely eliminating populations of moorhens, ducks and water voles have not materialized. Concern in Iceland that mink would have serious effects on birds has been allayed (Bengston, 1972) and destruction of stocks is likely to occur only when the mink have no alternative foods. Thus, concern is being voiced at the time of writing about a plan to set up a mink farm on Westray, one of the Orkney Islands and an important breeding site for seabirds, in view of the extinction of black guillemot and puffin colonies by mink in parts of Iceland and Sweden (Bourne, 1978).

American mink now coexist with both European mink and otters. They share the habitats and may share the same diet. It seems likely that the European mink may be driven out by the American mink through direct competition but probably it has no significant effect on otters. If there is competition for fish, the otter is the more adept hunter and has the edge on the mink which will be forced to hunt other quarry.

EUROPEAN POLECAT
(Mustela putorius)

The polecat has collected a worse reputation for raiding poultry runs than any of its relations, which can be traced in the origins of its name. The English name polecat is derived from *poulechat*, which is made up from the French words for chicken (*poule*) and cat (*chat*). A mid-Victorian writer describes the polecat as 'dreadfully destructive to the poultry' and tells of a polecat's burrow being stocked with a whole brood of ducklings recently lost by a farmer. At one time, the polecat had alternative names of fitchet or fitchew, from the old Dutch name, and foumart or foulemart, which are a contraction of foule-marten and refer to the polecat's habit of squirting an evil-smelling, milky fluid from the anal glands when alarmed. The pine marten was known as the sweetmart in comparison.

The European polecat is considerably larger than a stoat and is a little smaller than a pine marten, but has shorter legs than the latter. The body length is 31·5–45 centimetres, with a bottle-brush tail of 12·5–19 centimetres and weight 500–1,200 grams. The females are 40 per cent lighter and have narrower skulls. The dense underfur is pale yellow or tawny and is covered by long dark brown or black guard hairs. The coat is darkest on the head, tail, feet and underparts because the underfur is grey on these parts. On the rest of the body the guard hairs are loose so that the yellow underfur shows through to impart a brownish tinge and a yellow patchiness. As the polecat moves its coat changes colour through rearrangement of the two colour components. In one limited area of Wales there is a very rare 'red' variety in which the black of the coat is replaced by red or ginger. The winter coat is paler partly because the underfur is more white, as is the basal part of each guard hair, and also because the underfur is more dense and makes the guard hairs stand out. It would be interesting to know whether the paler winter coat has an adaptive value like the white coat of a stoat. The head has very distinctive patterning. There is a yellowish-white patch on each side of the head, between ear and eye, which, in the winter coat, tend to meet in the centre, and continue round the cheeks to join another patch on the muzzle. The ears are bordered also with yellowish-white. It would also be interesting to know whether the bold facial patterning is analogous to that of a badger. If the function is to warn off predators (p. 113), then we may wonder why badgers and polecats have facial patterns but wolverines and pine martens do not.

The polecat is found from the coast of the Mediterranean to the southern edge of the pine forests in Sweden and Finland and eastwards to the boundaries of pine forest and steppes in Russia (Corbet, 1966). It is absent from Ireland, Sicily, Sardinia and Corsica, European Turkey and the eastern shores of the Adriatic Sea. The habitat is very varied. It

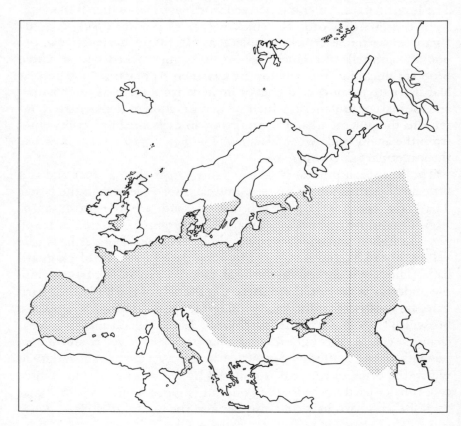

is found in woodland and marshes, on farmland and even in towns and large cities (Ognev (1931) says that polecats appear in Moscow), but low-lying, damp areas are preferred. The polecat has been persecuted for its fur, known as fitch, as well as for its carnivorous habits, and was once hunted with hounds in England, yet in most countries it has survived well. Only in Britain has the polecat's range been severely restricted. Once common—it was abundant around London until the mid-1800s—it survived until recently only in Wales, but it has now spread to adjoining counties of England. Unlike other predatory mammals and birds, the polecat found no refuge in the Scottish Highlands. The last Scottish polecats were trapped by 1907 whereas the species survived in parts of England for another three decades. A recent suggestion is that the polecat suffered more from the attentions of gamekeepers than did the pine marten because it prefers low ground, while the latter was able to retreat into highlands where it was not disturbed (Langley and Yalden, 1978).

The habits of wild polecats are not well known. They are solitary and nocturnal, and they presumably occupy home ranges like other mustelids. Faeces and urine are used for marking, as are the secretions of the anal glands, being used for this purpose rather than for defence as is sometimes supposed, although fluid is ejected when the polecat is alarmed. The polecat spends most of its time on the ground but can climb well. A captive polecat is said to have been able to climb paper-hung walls (Ognev, 1931).

The main food of polecats is mammals, particularly rabbits. Voles are favoured after rabbits and a quantity of birds, amphibians, reptiles and invertebrates are also taken (Walton, 1970). Frogs are eaten in large numbers when they are available; for example, polecats den up under peat hags in Tregaron bog, Wales, during spawning time (Owen, 1965).

Like the stoat, a polecat hunts mainly by smell. If a piece of meat is placed in the cage of a captive polecat, it approaches with rapid and seemingly random movements of the muzzle and there is little doubt that it does not notice the meat until about 30 centimetres away. Young polecats learn to recognize the odour of particular prey animals by smelling the food brought in by their mother. This fixes their food preferences for the rest of their lives. Although they can learn to hunt new kinds of prey, they always prefer to hunt animals familiar from their childhood diet. Apfelbach (1973) reared polecats on one kind of prey animal and found that when adult, they would ignore all other food.

The polecat follows the trail of its prey with its neck outstretched and the rest of the body lying so close to the ground as to be almost touching it. The spine must form an almost perfect straight line from one end to the other. As soon as the polecat stops, to reconnoitre or to investigate something, the body passes easily into a series of graceful curves. Once it has found its prey, the final attack is guided by sight and the position of the eyes and ears on the prey's head may help to orientate the bite to the neck (Wüsterhube, 1960). Biting the neck is an instinctive pattern of behaviour which appears in hand-reared polecats when four weeks old and is well aimed even though the kittens are still blind at this age (Poole, 1966).

The breeding habits of the polecat are at present largely a matter for conjecture. A rough mating, in which the male drags the female around by the neck, takes place in March and April. Release of the eggs from the ovaries is induced by copulation and the litter is born in May and June after a gestation of forty-two days (Corbet, 1966). There are four to six kittens in a litter, born in a nest of dry grasses among rocks or in a suitable hole such as a rabbit burrow or badger sett. Newborn kittens are naked and are blind for five weeks. The first coat is of sparse white hair and the adult marking appears at three months. A captive female carried meat to her kittens when their eyes were first opening, then would drag it out of the nest and drop it. Apparently this was a method of coaxing the kittens to learn how to follow a scent trail (Goethe, 1940). In hot weather Goethe's kittens lay as far apart as possible in the nest and their mother was seen to wet her belly in her water dish and then curl her body around the kittens to cool them. The kittens become independent at three months.

It used to be said that polecats bore two litters per year but a single litter is the rule except when the kittens fail to survive. Then, the female comes into heat again, mates and produces a second litter.

STEPPE POLECAT
(Mustela eversmanni)

Also called the Asian polecat or Siberian polecat, the steppe polecat is very much like the European polecat and is sometimes considered to be no more than a subspecies. In appearance, the steppe polecat differs from the European polecat in its paler head and body, in which the facial markings are less distinct. The legs and tail are dark brown. Body dimensions are smaller than those of the European polecat.

The steppe polecat ranges westwards from China and Mongolia to the Carpathians. Its European range includes Czechoslovakia, Hungary, Poland, Romania and Russia, north to a line running about 100 kilometres south of Moscow. Whereas the European polecat prefers woodland, the steppe polecat, as its name suggests, prefers open country such as semi-deserts and meadows. Consequently the two species stay apart where their geographical range overlaps and the steppe polecat is spreading northwards as forests are felled (Novikov, 1956), presumably at the expense of the European polecat.

Nothing is known of the social organization of steppe polecats and little about their behaviour except that they are more sociable and more diurnal than the European species and they live in burrows, where they may lie up for several days in bad winter weather. A steppe polecat may dig its own burrow or take over those of rodents, foxes and badgers. Rodent burrows may be enlarged and lengthened. It seems that a burrow occupied by a steppe polecat is readily identified by its evil smell, because a litter of uneaten scraps and unburied droppings accumulates around the mouth (Novikov, 1956). Surplus prey is cached, fifty ground squirrels having been found at one burrow.

The main food is rodents; mice and rats are most commonly taken over most of the polecat's range but ground squirrels and common hamsters predominate in the south, fish and amphibians becoming more important in the north. Birds and their eggs, reptiles and insects are taken only occasionally. Rodents are pursued down their own runs, which are widened as necessary to allow the passage of the polecat, while in winter the steppe polecat digs through the snow to capture mice and voles.

Although there is a difference in size between male and female steppe polecat, there is no difference in diet as in the pine marten (Yurgenson, 1947). The female is three-quarters the size of the male (585 grams compared with 812 grams) but she has proportionately larger muscles on the shoulders and jaws. These are the main muscles to be involved in handling prey so the smaller female is as capable of handling marmots larger than herself, as is the male.

Mating takes place in February or March and the kittens are born some 36–40 days later. The litter is usually of eight to eleven but eighteen is known. The kittens develop more rapidly than do those of the European polecat. The eyes open at two weeks instead of five and weaning is accomplished in six weeks. At about this time, the kittens sally forth in a pack to hunt ground squirrels and they become independent at the end of the summer.

MARBLED POLECAT
(*Vormela peregusna*)

Sufficiently different from the other polecats to deserve a separate genus, *Vormela*, the marbled polecat is a striking animal. Marbled refers to the variegated coat. The back and bushy tail are mottled dark brown and cream; the rest of the body is dark brown, including the underparts, but there is a white band across the forehead and white muzzle leaving a dark 'robber mask' over the eyes. There is considerable variation between individuals. It is smaller than a European polecat, the head and body length is 31–38 centimetres, tail length 15–16·5 centimetres. The claws are longer than in other polecats.

The marbled polecat ranges westwards from the Gobi Desert in Mongolia across the Russian steppes to Romania and Bulgaria. Its typical habitat is open steppes, scrub and cultivated land, but it also lives in forests. In Europe, it is found principally on virgin steppe and old fallow ground but it also inhabits arable land and may enter buildings (Novikov, 1956). It is nocturnal for the most part and lies up in a burrow excavated by itself with its long claws or abandoned by another animal. One burrow occupied by a family group extended 5·5 metres, just below the surface, then dropped vertically for at least one metre (Ognev, 1931).

The coloration and behaviour of the marbled polecat lends support for the supposition, discussed already in relation to the badger (p. 113) and the European polecat (p. 98), that striking patterns of body colour are a warning to predators that the wearer should be left alone. The anal glands of the marbled polecat secrete an evil-smelling fluid and, when threatened, the teeth are bared, the hair of the body and tail bristles and the tail is thrown forward in a warning display. Moreover,

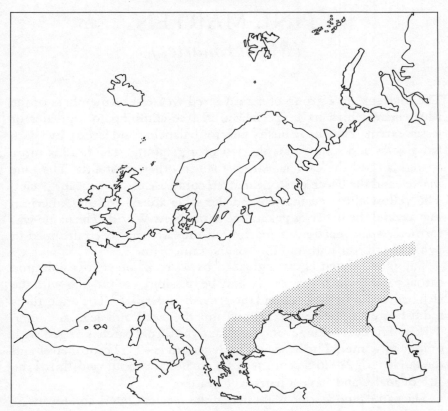

marbled polecats are very bold, as if aware that their effluvia convey immunity from attack.

In essence, the habits of the marbled polecat are little different from other polecats, in so far as these are known. When there are trees it will climb but the marbled polecat is basically a terrestrial hunter. Its food is mainly small rodents such as will be caught on the steppe: voles, gerbils, ground squirrels, hamsters, mice and young hares, as well as birds, reptiles and amphibians. Poultry killing has led to persecution in some places and there is a small market for marbled polecat pelts in eastern Europe.

The gestation is said to be nine weeks (Walker, 1964), the four to eight kittens being born in a nest of leaves in the burrow in March or April. They stay with their mother at least until June (Novikov, 1956).

PINE MARTEN
(Martes martes)

The martens are a group of six cat-sized weasel-like members of the Mustelidae which have specialized in tree-climbing to a greater or lesser extent. They have bushy tails for balancing and large paws with hairy soles and semi-retractile claws for gripping. The head is more triangular and the ears are larger than in other mustelids. The pine marten and the beech or stone marten comprise the European species. The yellow-throated marten and the sable are Asian; the American marten and the fisher or pekan live in the New World. The name was formerly spelt martren or martin but the latter has been dropped to avoid confusion with the bird of the same name.

The pine marten is distinguished by a cream or yellow coloured throat patch (which very rarely may be missing), contrasting with the rich brown fur of the back and the greyish underparts. The underfur is reddish-grey tipped with yellow and the head and legs are often darker. The coat is longer and finer, and the throat patch is tinged with orange in winter. The head and body length is 42–52 centimetres and the tail length 22–26·5 centimetres. Females are about two-thirds the size of males and have a narrower head.

The name pine marten is apt in that the species prefers coniferous or mixed woodland to deciduous woods but it also lives in open, treeless country. Its range covers most of Europe, except for most of the Iberian Peninsula, large parts of the Low Countries and north-west France, Greece, parts of Finland and most of the British Isles. It also inhabits a broad belt of Asia extending northwards to the edge of the forests and across to the Gulf of Ob and Lake Balkhash. Although once abundant, the species declined rapidly as preservation came into vogue. Gin traps set out for rabbits and foxes proved to be very destructive of pine martens. Its rich fur was also greatly prized and 'hunting the mart' by gangs of men armed with sticks and stones was a popular mediaeval pastime. In Britain, the pine marten survives only in the wilder parts of northern England, Scotland and Wales, but it seems to be more widespread in Ireland.

Perhaps because its numbers are now very low, the pine marten is more difficult to track down than the almost proverbially elusive otter. In Scotland, it was trapped out during the nineteenth century until it survived only in the remote north-west. Since World War II its

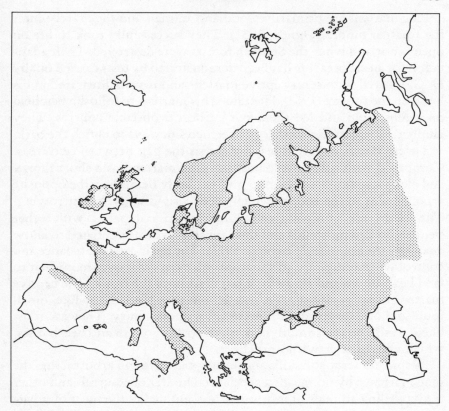

range in Scotland has expanded mainly, it would seem, through reduced persecution. Many of the records are widely scattered. Typically, a marten is caught in a fox trap at a considerable distance from the nearest previously known record. Then no more are caught in that area for many years.

The pine marten has always been considered to be a nomad. It was said to move from place to place in search of hunting grounds and to cover vast distances. The same was said of the otter, that it would occupy a lake or stretch of river for a time, then move to the next, until more careful observation showed that each otter had a home range, albeit a large one, to which it restricted its activities. The pine marten organizes its life in the same way. Each marten has a large home range which it marks by smearing anal gland secretions on the ground or along boughs. The Scottish martens trapped outside the well-established enclaves must be the same kind of dispersing wanderers as are found in otter, stoat and weasel populations.

Pine martens are primarily woodland animals and they rarely enter the Russian tundra (Ognev, 1931). They successfully took to life on open moorland when the Scottish forests were destroyed. Their adaptation for an arboreal life has been demonstrated by tests carried out by H. G. Hurrell with semi-captive martens and is well illustrated in his book on the subject (1963). He trained his martens to follow a bait held on a long pole and led them over arboreal obstacle courses. They climbed like squirrels, with legs stretched sideways to clutch the bark, and were just as expert at leaping across the gap between two trees. Compared with other mustelids, the pine marten has a short thorax and elongated lumbar region to give the body flexibility, the opposite arrangement to the badger which needs a rigid body for burrowing. The limbs are comparatively long, when compared with other slender-bodied mustelids and their muscles are differentiated to allow dexterity (Sokolov and Sokolov, 1971). The tail is used for balance and control when jumping and a marten will even twist over in mid-air to land head-down and ready for a fast descent of the trunk. The captive martens occasionally slipped and fell but they landed, cat-like, on all fours and one survived a 20 metre fall without injury. They are most likely to slip on the smooth bark on the trunks of ash and beech trees, which allow the claws no purchase.

Despite its acrobatic skills and an adaptation for an arboreal life, the pine marten is by no means a specialized hunter of squirrels and other tree-dwelling animals. It is omnivorous and makes the most of whatever prey is abundant (Lockie, 1964 and Yurgenson, 1947). Where squirrels are abundant, they are eaten, but on the Scottish moors for example, where there are no squirrels, other food is sought. Lockie estimated that over half the pine marten's diet, by weight, in this habitat was made up of small rodents (Lockie, 1961). As with foxes (p. 33) and weasels (p. 80) field voles were preferred to other rodents. They appeared in nine out of ten marten droppings when collected by Lockie. No more than one in ten rodents caught in traps were field voles. If we can assume that the traps catch rodents without any bias towards one or other species, then the martens are catching field voles selectively.

The next most popular items of food are small birds and, showing how omnivorous martens are, fruit and pine cones. Large birds up to the size of wood pigeon and grouse are caught on occasion but smaller species are commonly hunted, especially in winter, including chaffinches, pipits, blue tits and wrens; and eggs are also eaten. Fruit is very popular when in season; bilberries, rowanberries and wild raspberries are gorged and gardens are sometimes raided for cultivated fruit.

Deer carrion is eaten in Scotland and Osgood Mackenzie (1924) records an attack on a sheep. The sheep was killed by the marten which died in turn as it was squashed against a boulder under the dying sheep. Young roe deer are sometimes attacked in Russia and Ognev (1931) even records a successful attack on a wild cat. Lesser food includes hares, the caterpillars and pupae of oak eggar moths, and fish. Wasps' nests are dug out and one of Hurrell's tame martens caught bumblebees by knocking them out of the air with a swipe of the paw. He also saw his martens catch cockchafers in oak trees and clean slugs of slime by rolling them on the ground before eating them. The smaller female pine martens are unable to catch such large prey as the males can. In Russia, they eat small birds, wasps, voles and pine cones whereas males catch hares and gamebirds such as capercaillie and grouse (Yurgenson, 1947).

Pine martens are mainly nocturnal but can be seen abroad by daylight, particularly at dawn and dusk. Each has a nest or lair which may be in a hollow tree, in a rock cleft or under boulders. Abandoned nests of squirrel, goshawks and others are also utilized.

The mating season is in summer, from June to August. There are two or three periods of oestrus, at two-day intervals, during which the pair stay together. Mating sometimes occurs in trees but more often on the ground. Ovulation is induced by copulation as in some other mammals, e.g. rabbit and mink, rather than occurring spontaneously at oestrus. The penis is provided with spines, perhaps to ensure adequate stimulation for ovulation to follow (Ewer, 1973). Implantation is delayed until February so that birth takes place in the following April after a eight- to nine-month gestation. The two to six kittens are born with sparse whitish hair. Their eyes open at two weeks and the first steps outside the nest are taken when about eight and a half weeks old. The tail lacks the bush at this stage and the kittens are ill at ease. They cling tightly to the trunk or bough as if frightened of heights. When lost, a kitten will call with a sound like tearing paper (Hurrell, 1963). The family breaks up in late summer by which time it is too late for the female to mate again and so she breeds only once in two years.

BEECH MARTEN
(Martes foina)

Also called the stone marten, the beech marten is very similar to the pine marten in appearance and habits. The main difference in appearance is that the beech marten's throat is white and the pine marten's cream or yellow. The muzzle is shorter and broader, the feet are less well haired and the ears are narrower and shorter. The body is heavier, although slightly smaller, and the appearance is generally more squat. Head and body length is 42–48 centimetres, tail 23–26 centimetres and weight is 1·3–2·3 kilograms.

The beech marten overlaps the range of the pine marten over much of Europe, but is generally found farther south. It is absent from Scandinavia and northern Russia but spreads through the Iberian and

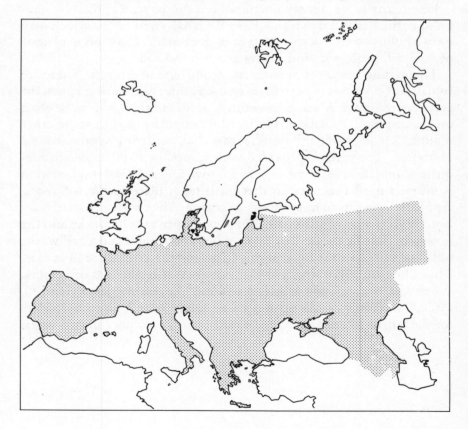

Balkan regions, as well as through Asia Minor to northern India. It is missing from the British Isles but otherwise extends from the western seaboard of Europe across to Mongolia where it is replaced in the Far East by the yellow-throated marten.

The alternative name of stone marten is more apt. Although it inhabits forests, the beech marten is less attached to them than the pine marten, and often frequents open country. In the Mediterranean region beech martens live on rocky slopes and mountain ravines. Elsewhere, quarries are occupied and they are not afraid to enter buildings. Ognev (1931) records that beech martens have lived in the attic of the university building in Kiev and the species is called *Hausmarder* (house marten) in German. Although a good climber, the beech marten is more terrestrial than the pine marten even in woodland habitats. When chased, it prefers to run to cover rather than climb a tree (Novikov, 1956).

The behaviour of the beech marten has not been studied in detail but where known, the habits are similar to those of the pine marten. The usual den is a hollow tree or rock crevice but attics and even old storks' nests are employed. Mating takes place in summer, during two or three periods of oestrus, and there is delayed implantation whose length is said to vary with geographical location. In European Russia, gestation lasts for 236–274 days, so that the young are born in the spring following mating. The litter consists of three or four kittens.

The diet is also largely similar to that of the pine marten. Voles and mice are the chief prey, followed by insects and birds, with the occasional shrew, mole or dormouse. Fruit becomes an important part of the diet in autumn. Novikov records that in the Transcaucasus, 59 per cent of beech martens examined had eaten the fruit of wayfaring trees, and 11 per cent had fed on haws and pears. In Poland, fruit and seeds continue to be important through the winter until they become scarce. Small mammals and birds make up the bulk of the diet in spring and summer (Ryszowski *et al.*, 1971).

WOLVERINE
(Gulo gulo)

The largest of the mustelids, the wolverine has been described as a bear-like 'super weasel'. It has a reputation for ferocity, strength and cunning but the alternative name of glutton seems not to be fully justified, even though it will attack animals far larger than itself and eat almost anything. It will kill deer but the statement which used to be frequently repeated, that a wolverine does not stop eating until there is nothing left completely disregards the size of its stomach in relation to the prey's body. Nevertheless, the wolverine's teeth and jaws are strong enough to reduce the limb bones of deer to splinters.

One commentator has pointed out that one of the wonders of the world, for which there is no explanation, is the human capacity for exaggerating until a basic fact is obscured by romantic decorations. As will be seen, the wolverine is a remarkable animal, but this has not been enough for some. There is, for example, an old belief, which persisted until not so long ago, which concerned the wolverine's stratagem for catching deer. The wolverine was said to climb a tree carrying a mouthful of moss, and when a deer approached the wolverine would let the moss fall. Should the deer stop and eat it, the wolverine would drop on its back, fix itself firmly between the antlers and tear out the victim's eyes. Either from the pain or to rid itself of its tormentor, the deer would bang its head against a tree until it fell dead.

Although heavy-bodied and bear-like, the shape of its head, with pointed snout and small ears, marks the wolverine unmistakably as a mustelid. It weighs 9–30 kilograms and measures 70–82·5 centimetres head and body length, with a 12·5–15 centimetres bushy tail, males being larger than females. The coat is dark brown with a pale ring around the face and a yellowish streak running down the sides of the body. The soles of the feet are covered in hair, like those of a polar bear, the hair acting apparently as an insulation from snow and ice. Wolverines have long been trapped for their fur, which is traditionally used for trimming the hoods of parkas. When the hood is pulled tight, the long hairs form a mesh which traps a protective cushion of warm air in front of the face. The particular advantage of wolverine fur is that ice condensed on it from the wearer's breath can be easily shaken off, whereas wolf fur has to be thawed to remove rime and so becomes soggy.

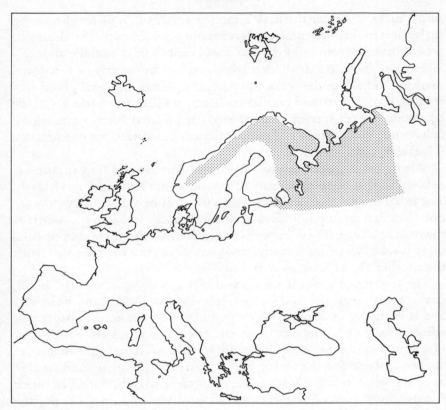

Wolverines are found in tundra zones of both Old and New Worlds, and in Scandinavia they live in mountainous country, both above and below the tree line. The European distribution covers the Kjolen mountain range in Norway and Sweden, except in the part south of Bergen, around the north and east of Finland and into parts of the Baltic States. Occasional wanderers are found well to the south of the normal range. Wolverines have for long been persecuted, either for their fur, or because they steal animals from traps and raid reindeer herds. The result is that they have been eliminated from many places and are rare elsewhere.

Wolverines are largely nocturnal and are solitary except when the sexes associate for mating and when females are accompanied by their young. Females have ranges lying within those of the male and a pair will occasionally walk together for some distance. The male's range is huge and can cover as much as 2,000 square kilometres (Knott, 1960). The females range is smaller and several may live within the range of a single male. Whether the male's range overlaps that of neighbouring males or is exclusively defended seems not to be known. The sheer size would make it seem unlikely that a wolverine could exclude all interlopers but these are dissuaded from settling by means of territorial marking. Scent, in the form of urine, droppings and anal gland secretions, is placed around regular trails and at kills, and marks are made by biting pieces out of tree trunks, an unusual form of marking behaviour. Pine, spruce and juniper trees are preferred for this practice (Haglund, 1966).

Mating takes place between April and August and implantation is delayed so that the birth occurs in the following February and March. The usual litter is two or three, with a maximum of four. The cubs are born in a den among boulders or in a hole excavated in a snowdrift at the base of a crag. They are well-furred at birth and their eyes open at three weeks. Weaning is complete at ten weeks but the cubs stay with the mother for as long as two years.

The wolverine's diet is varied and includes deer, lemmings, hares, foxes, birds' eggs, fish and berries. It is not a skilful hunter, however, and it hunts by quartering its range until it comes across something edible. Deer, up to the size of an elk, foxes and hares are chased, but Haglund found that most chases are unsuccessful and that success in winter depended on the wolverine's broad paws giving it an advantage over the prey in soft snow. Several reindeer may be killed in quick succession if they cannot escape, old or diseased animals being the most likely victims. If the snow is too soft, the wolverine generally will not bother to give chase, but some pursuits on soft snow cover upwards of 5 kilometres. Large animals, such as reindeer, are killed by the wolverine jumping on the victim's back and biting the neck or back.

Much of the wolverine's food, at least in winter, consists of carrion. Haglund noticed tracks of wolverines following those of lynxes and concluded that the mustelid was deliberately following the cat in the hope of a free meal. In former times, when they were more numerous, wolves were also unwitting providers of wolverines. The wolverine's reputation for ferocity is well founded in that it can drive lynxes and wolves from their prey, by raising the hair of its back, baring its teeth and growling in an awe-inspiring display of threat.

Surplus food is cached by being buried in snow, hidden in crevices or among boulders and carried into trees. Another testimony to the wolverine's strength is its habit of dragging its prey over great distances, sometimes up to 10 kilometres before caching it.

1   Paternal solicitude: a dog-fox brings food for the vixen and cubs

2   The thick, white coat of the Arctic fox is an aid to survival in the harshest climates

4 The golden jackal thrives in south-east Europe and sometimes enters towns in search
of food

3 (*Left*) The wolf, progenitor of the domestic dog, survives only in the wildest places

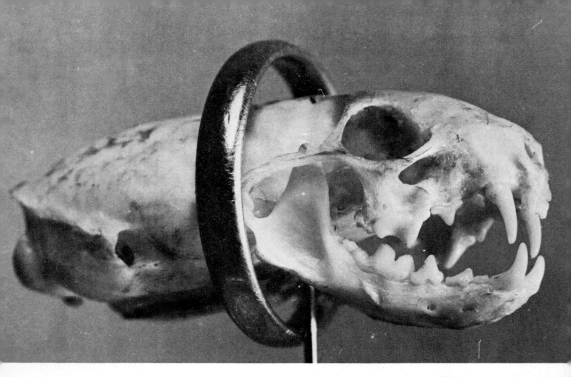

5 Tradition says that a weasel can pass through a wedding ring

6 A weasel looking up to show the two characteristic brown spots on the throat

7 The European mink has disappeared from many parts of its former range but the very similar American mink is spreading rapidly

8 The agility of a polecat is caught by the camera as it leaps across a stream

9 Another hard-pressed carnivore, the wolverine is persecuted for attacks on livestock

10   The otter's attacks on fish make it unpopular with anglers

11 Belying a reputation for ferocity, a tranquil view of a captive wild cat

12 The raccoon was introduced to Europe as a fur-bearer and thrived to become Europe's sole representative of the family Procyonidae

BADGER
(Meles meles)

Badgers are heavy-bodied mustelids with some of the characteristics of a bear. These include broad, plantigrade feet with strong claws, and an omnivorous diet. They are nocturnal, coming out at dusk, returning at dawn and are rarely seen by day. Yet despite retiring ways, badgers are favourites among naturalists because their habit of living in traditional burrows, called sets or setts, makes them relatively easy to watch. Most of our information about badgers has come from watching as they emerge from the sett and by following their well-worn trails (Neal, 1969), but more detailed studies based on the movements and activities of marked individuals are now being made.

The badger has a broad body, tapering from the hindquarters to a relatively small head, and short strong legs. From the tip of the short tail to the snout, a badger measures up to 75–90 centimetres. Weight ranges from about 10 to 22 kilograms and varies through the year, being greatest when the winter store of fat has been laid down. Females, or sows, are a little smaller than males, or boars, and can be distinguished by a narrower face. The body colour is grey but close examination shows the coat to be made up of hairs which are pale at tip and root, and dark in the centre. Unusual variations include albinos, almost black melanics and reddish erythristic forms. In contrast to diurnal animals, whose pale underparts are said to be an aid to camouflage, the underparts of the nocturnal badger are darker than the back. The head is distinctive with its broad black and white striping. At one time there was a theory that the stripes are a camouflage to break up the badger's outline as it wanders through moonlight-dappled woodland. This notion does not stand up to critical observation and a more likely explanation is the opposite—that the badger is designed to look conspicuous, as suggested by Pocock as long ago as 1911, to advertise that it is not an animal to be trifled with. It is a stern adversary, hence the onetime popularity of badger-baiting with dogs. The bold stripes warn that an adversary runs the risk of a severe mauling by the badger's claws and teeth. The eyes lie in the black stripes and it has been suggested that other animals possess an eyestripe either to disguise the vulnerable eye or to act as an anti-dazzle device. A further deterrent to a predator, possibly a lynx or wolf, is the badger's ability to erect the hair on the body to look twice its actual size.

The skeleton is stoutly constructed and modified for a burrowing way of life (Sokolov and Sokolov, 1970). The thoracic part of the vertebral column is lengthened at the expense of the lumbar region, while the pelvis and its associated muscles are well developed to lock the hindlimbs against the burrow walls when digging. The hands and feet are short and armed with strong claws for scraping earth. The forelegs are extremely muscular and the handbones are well developed to give a broad palm for shovelling but the humerus is not curved as in other digging animals (Ondrias, 1961).

The strength of the skull is remarkable. The sagittal crest on the top of the skull enlarges the area for attachment of the massive jaw muscles (see p. 14) and the jaws are held in place by flanges. In most animals the jaws fall away from the skull when the flesh is removed, but the badger's jaws cannot be removed without smashing the bone. This would seem to be an unnecessary strengthening as badgers eat mainly soft food but it can be seen as a further development of the peculiar system found in other mustelids where the muscles work to the best advantage when the jaws are nearly closed. Their force tends to dislocate the jaw and the hinge has to be modified to hold it in place.

Some nocturnal animals have large eyes to enable them to see well in dim light, but badgers have poor vision and rely on other senses. Even when abroad on a fine summer's evening, when the sun has barely set, badgers may come within a metre or so of a watcher before another sense warns them of danger. A badger can, however, pick out a movement against a skyline. Its hearing is acute but the sense of smell is paramount. On emerging from the sett, the badger points its nose at the sky and samples the air for tell-tale scents before starting on the night's activities.

Communication involves the principal senses of smell and hearing. Like other mustelids, the badger is equipped with subcaudal musk glands under the base of the tail whose prime function is scent-marking. By squatting against boulders and tree roots and depositing a smear of musk, the badger leaves its own personal imprint, which helps to create a familiar atmosphere. Ewer's (1968) tame badger would become agitated if brought into a strange room but calmed down when presented with an object which it had previously marked. Phil Drabble's (1969) semi-wild badger set a trail of marks when foraging along new ground and used them to navigate home. The scent-marks also tell other badgers of the marker's presence and the boar marks the sow before mating. Drabble (1969) suggests that mutual grooming by members of a sett before emergence helps to promote familiarity. Musk is also emitted into the air to leave a rank odour when badgers are excited by playing or courtship. Secretions from the anal glands have a rather different function. They have a powerful musky odour and are particularly noticeable in the spring and seem to be a means of marking territory. A sudden scare causes the badger to void its anal glands. Perhaps a development of this last habit led to another mustelid, the skunk, to use its musk glands for defence.

A variety of calls are given at different times. The boar makes a prolonged low purring which has sexual significance and is heard before, during and after mating. A rather similar sound is employed by the sow towards her cubs, as well as one described as sounding like a moorhen and used either as a warning or during courtship. Barking and growling are used as threats between badgers. The cubs squeak and yap noisily but the loudest noise is the badger's yell. Eerie and scaring, the yell's significance is not known but it has been heard from a snared badger.

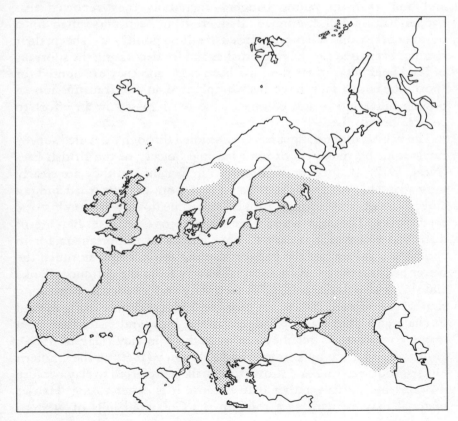

The European badger should strictly be called the Eurasian badger because its range extends from Ireland and Portugal to the eastern seaboard of China and Korea. In Scandinavia, the European badger penetrates the Arctic Circle but its northern limit moves southwards to avoid the Central Siberian Plateau and the mountain ranges of north-east Asia. To the south, the badger is limited by the Mediterranean Sea and the Himalayas. In general, badgers are found wherever there is wooded and hilly country. Thus, they are common in Germany and Scandinavia but very rare in central Holland where there is simply nowhere suitable for them to live in the flat peaty country. The sprawl of towns has reduced the area where badgers can live but they survive in parks and waste ground surprisingly near the centre of cities like London. However, the experience of the Dutch is that isolated pockets of badgers eventually die out. More regrettable is the finding of a survey in the Netherlands (van Wijngaarden and van de Peppel, 1964) that legal protection had not prevented farmers, gamekeepers and hunters from killing badgers. Incredibly they reported that badger-baiting still continued. Badgers have been persecuted for a number of reasons. They are accused of killing poultry and sheep; their hair has been used for shaving brushes and for decorating the sporrans of Highland Scots; their flesh has been eaten and they are hunted for sport. Recently, they have been implicated in the transmission of bovine tuberculosis and colonies have been gassed in an effort to control the disease.

The habitat of the badger has been studied through a detailed survey carried out by members of the Mammal Society of the British Isles (Neal, 1972). Badgers are widespread in Britain and they are clearly very adaptable with their setts being found on sea cliffs, road and rail embankments and cuttings, rubbish and mine tips, even in prehistoric earthworks. The man-made sites are attractive because they are of disturbed soil which is easy to dig. This is an important factor in choosing a site and, if the badgers can dig into a bank, so much the better because the excavated soil falls away from the entrance. Banks and slopes also drain well and in Germany and Sweden, south-facing slopes are preferred. Whenever possible, badgers choose to dig in sand or chalk and shun clay. A badger can dig into sand faster than men trying to unearth it but they can overtake it in clay (Davies, 1936). However, a survey by Dunwell and Killingley (1969) in the Chiltern Hills of southern England shows that chalk is preferred to clay because it drains more easily rather than because it is easier to dig. Harder material may be used as a roof. In the Cotswold hills of western

England, badgers burrow into soft sandstone just under a stratum of limestone or into clay with marlstone above. Badgers have even used the surface of a main road as the roof of a sett. A strong roof is a decided asset as setts sometimes cave in, particularly where heavy farm machinery passes overhead!

Badgers are not found at any great altitude, as high country is unlikely to provide much food, while they avoid low-lying country which is liable to flooding. An exception is Neal's finding (1962) of a badger nest, complete with cubs, in a hedge on easily flooded, difficult to excavate, country in western England, and of another under a pile of planks in a disused farmhouse. In Argyll, on the west coast of the Scottish Highlands, setts may be found on the shore a couple of metres above the high tide mark as well as 200 metres high on the moors, where the badgers lie up in cairns of rocks or burrow into hummocks of boulder clay. Elsewhere, setts have been found at up to 500 metres.

Cover is an important consideration in siting a sett because the badger likes to emerge unseen, and even apparently open sites may have cover in the form of bracken or nettles in summer. Deciduous woods and copses are preferred to coniferous plantations, which lack ground cover, although the planting of conifers on otherwise open moors in Scotland has aided the badger. Setts are distributed along the edges of woods so the badgers have only a short walk to the feeding grounds in the fields, and, where woods are missing, hedges may be used. The sett also creates its own cover for well-established setts are frequently surrounded by stinging nettles and elders, two plants which thrive on disturbed, well-fertilized ground, and which are left to grow while other species are eaten by the badgers.

The badger's sett is unmistakable, and easily distinguished from burrows of fox and rabbit by the huge pile of soil that attests to the size of the excavation. At an old sett, the pile may be grassed over whilst a conspicuous gash of yellow or white shows where sand or chalk has been freshly ejected. Some setts are of great antiquity and place names such as Brockholes, Brockenhurst and Brockhampton in England, Loch an Bruic in Scotland, and Dassenbosch (meaning 'badger wood') in the Netherlands, to name but a few, indicate centuries-old allegiance to a particular site. Old setts may riddle 5,000 square metres of ground and can have over fifty entrances. Such extremes are unusual and there are usually fewer than half a dozen entrances. One or more will be favoured at any time and side entrances are used as bolt-holes.

The sett is excavated without plan and generations of badgers add new workings so that there may be two or three storeys and several hundred metres of tunnel in exceptionally large setts. At intervals there are breeding and sleeping chambers which contain bedding. Breeding chambers are often enlarged side tunnels up to 60 centimetres high. Ernest Neal has pointed out that they are often well protected by a solid roof such as a boulder or the stout roots of a tree.

Without keeping watch to see the badgers emerge, it is still possible to determine, with reasonable accuracy whether a sett is occupied. The chambers are well supplied with bedding which is thrown out at intervals, mainly from September to December in preparation for winter and from February to March before and after the cubs are born. In Highland Scotland, the harvest moon used to be called 'Brock's Moon' because its light shone on badgers as they collected winter bedding. Huge masses of old bedding are stacked outside the sett while trails of litter show that new material has been brought in. Dead leaves, bracken, grass, moss and other plants are used. The badger gathers a bundle with its forepaws, heaps it under its chin and shuffles backwards to the sett, or noses the bundle forwards.

Linking the sett entrances and radiating across the surrounding country will be a number of paths, cleared of vegetation and beaten so hard by the pressure of countless paws that they survive long after the sett is deserted. At first sight, these tracks could be mistaken for human footpaths until they run under low branches and through hedges. Where the badgers have passed under fences, telltale strands of black and white hair are caught in barbed wire.

Footprints are left in damp soil, particularly where the badgers go to drink. They are unmistakable, with the typical five toes of the mustelid family and an extremely broad pad. The forepaws of a male may be 5 centimetres across, while the hindfeet are about 4·5 centimetres and those of a female are slightly smaller. If the claws register, it can be seen that they are large and sharp to serve as digging tools. They are kept sharp by being raked on the trunk of a special scratching tree near the sett entrance. Like a cat, the badger stretches on its hind legs, reaches up with the forepaws and drags them down.

Fresh mud on the scratching tree is a good indicator that badgers are in occupation but perhaps the most reliable sign is newly used dung pits. Badgers are cleanly animals, depositing their droppings in small pits in the manner of a cat, but not covering them. Some dung pits are sited within a few metres of the sett or below ground in a side chamber of the sett, but others are found at a considerable distance and have a role in the social life of the badger community which will be described later. Ernest Neal says that a careful watch for the appearance of unusually small droppings shows when the cubs have come above ground for the first time.

The regular habits of the badger extend to the time that it emerges from the sett for the night's activity. Within limits, emergence occurs at the same time each night and gets later as the nights get shorter. In winter the badgers come out about two hours after sunset but, in midsummer, when the time for nocturnal activity is severely limited, they are active shortly after sunset. Light intensity is most likely to be the main regulator. Cloudy nights bring the badgers out earlier (Lloyd, J. R., 1968) and warm dry nights encourage early activity, particularly in winter, when emergence is more erratic, and during very hard weather several nights may elapse without the badgers stirring. The presence of cubs delays emergence and unmolested badgers may come out before sunset or, rarely, they may sleep above surface in dense undergrowth. In populated areas badgers are only abroad on the darkest nights.

There used to be a controversy over whether badgers spend the winter in hibernation. But as they emerge in midwinter, to leave footprints in the snow, and give birth to their cubs in February it is reasonable to conclude that they do not hibernate in the strict sense, in which the metabolism runs down and the animal becomes effectively cold-blooded. Rather, badgers enter periods of dormancy in which the body functions slow no more than in normal sleep. In the cold Russian winter dormancy is more complete and badgers will not emerge even during warm spells (Novikov, 1956).

It is significant that the badgers, bears and the raccoon dog are the only carnivores which enter a state approaching hibernation. All are omnivores and Ewer (1973) points out that true hibernation is the prerogative of mammals such as plant- and insect-eaters whose food supply disappears in winter after an autumnal abundance. Hibernation allows an economic use of stored food but carnivores cannot lay up sufficient supplies to hibernate because their prey is not particularly abundant in the autumn. The exceptions are the three omnivorous members of the Carnivora which gorge on berries and the like in autumn, and so can indulge in prolonged inactivity. It may be that in the most northern parts of its range, the badger enters true hibernation and exists on its store of fat. Neal (1948) considers that, when the climate was colder, the badger was wholly a hibernating animal and that it reverts towards its old habits in severe winters. As it is, badgers put on weight in autumn and can exist without eating during spells of bad weather. The badger's winter habits are not wholly resolved and it would be interesting to monitor body temperature and heart beat of sleeping badgers as has been done with denned-up bears.

It could be said that badgers eat anything edible. They are true omnivores. The molar teeth are large and more flattened than in other mustelids, as are the carnassial teeth, both being adaptations for crushing fibrous plant food rather than slicing flesh and bones. The great length of the intestine is also an adaptation for plant-eating. The ratio of intestine to body length is 8·5, almost the same as in bears, as compared with 27·0 in herbivorous sheep and 2·9 in carnivorous foxes. Studies of stomach contents and droppings by Neal (1948) and Skoog (1970) reveal an impressive list of plant and animal food. As diets vary with time and place and, perhaps, with individual preference, it is not surprising that descriptions of the badger's feeding habits are sometimes conflicting. In his study of badgers in Sweden, Skoog distinguishes between primary and secondary foods. Primary foods include earthworms, insects, plants and mammals. Secondary foods include birds, molluscs, carrion and other odd items.

In many places earthworms are the most important food, though less so in summer when other items are abundant. The common *Lumbricus terrestris* is most commonly taken. Insects come a close second to earthworms and they are taken most frequently during the summer. Wasps and bees are favourites despite their stings which the badger braves when it raids their nests. Scattering the comb over the ground, the badger devours honey, larvae and even adults. Not surprisingly, it is the ground-nesting species of bees and wasps which are taken and the same applies to beetles. Dor beetles and chafers are sought when they are common and many other insects are eaten as the opportunity arises, caterpillars and leatherjackets in abundance, and even ants which must take considerable effort to collect.

Small mammals are not eaten very often but, as one rabbit or even a mouse has the nutritive value of many worms and insects, they are an important food. Most of the victims are juveniles snatched from the nest. The common rodent prey is that favourite of all predators, the short-tailed vole, but rats are killed, and a few shrews, rabbits and hares. Even moles and hedgehogs are not safe. Most of the moles are probably caught as juveniles dispersing over the surface, but badgers have been recorded as digging up their runs. Hedgehogs are despatched and all that is left is the cleaned skin with the spines intact. It would be interesting to know whether the badger has a special ploy for dealing with a curled-up hedgehog or whether it is as heedless of sharp spines as it is of the stings of bees and wasps. The former possibility is indicated by the finding of very few spines in badger stomachs after they had eaten hedgehogs.

Plant food is most important in the autumn when fruits are ripe and the badger can feed well to lay up a store of fat against the winter. Bilberries and wild raspberries, windfall apples, blackberries and acorns are eaten in season, while oats and wheat crops are invaded if the sett is near arable land. Bluebell bulbs and other storage organs such as rhizomes are also favoured. The enormous appetite for berries is shown by the fact that 800 bilberries were recovered from one badger's stomach. Neal has found that badgers sometimes stripped trees of their bark and licked up the sap. Attacks of this kind are made in spring and usually on beeches and sycamores.

Of secondary foods listed by Skoog, birds form probably the largest category, particularly during the breeding season when eggs and young are an easy prey. While badgers are often blamed for losses among gamebirds there is no real evidence to support such a notion, although it must be presumed that a badger occasionally comes across a nest and does not disdain to devour the contents. Badgers are also accused of killing poultry but there is reliable evidence for very few attacks. Howard Lancum (in Neal, 1948) found proof of badgers being the culprit in two out of one hundred cases of poultry killing. Similarly, sheep–killing is a very rare occurrence. In both cases, foxes must shoulder much of the blame, especially when they have carried corpses back to the badger's sett where they are in residence. Badgers very rarely carry food home. Blame for killing has also been attached to the badger from its habit of feeding on carrion. Most adult birds, hares and rabbits that are eaten by badgers are already dead from other causes. Badgers savage prey such as poultry, inflicting grievous lacerations; and a deer carcase horribly savaged is a sure sign of a badger feeding on carrion.

Among the lesser items taken by badgers are slugs and snails, both easy to find on moist warm nights, and a few reptiles and amphibians. In Denmark badgers eat huge numbers of tiny frogs as they emerge from ponds. This by no means exhausts the list of badger food, for badgers even search seashores for mussels or cockles thrown up by storms.

An important point to note is that this wide diet is composed largely of food found on or under the ground. The badger forages by smell, collecting stationary items rather than hunting down moving prey. It is not surprising that a digging animal should find so much of its food underground. We have seen that it digs out rodents and perhaps moles, most likely taken from nests, while many earthworms, chafer grubs and leatherjackets are found by rooting with the snout, so that the ground becomes peppered with small conical holes. Wasp nests and the stops of rabbits, where the young are hidden in apparent safety, are dug out from above. But badgers do not keep their noses slavishly to the ground. They reach up to pluck blackberries. One was seen to climb currant bushes and Ernest Neal has photographed a badger shinning a tree trunk, apparently after tree-climbing slugs.

Descriptions of the reproduction of the badger were rather ill-informed until Ernest Neal (1948) made his systematic observations. Even now the whole story of the birth of badgers is not clear. The time of birth used to be a mystery and it was even more difficult to discover the mating season of this shy, nocturnal, burrow-dwelling animal. As badgers tend to live socially (see page 124), there is no obvious time for the sexes to associate but communal life is more pronounced in July, August and early September. However, mating occurs from February to October. There are two peaks of mating behaviour, February–May and July–September, with each sow coming on heat for about five days.

Mating can be quite casual and without preliminaries. The boar approaches the sow, they sniff and mark each other with musk and may groom each other (Paget and Middleton, 1974). Then the boar mounts, maintaining his position by biting the sow's ear. At other times there can be a deal of play. In many mammals the courtship period is a time when adults revert to a behaviour more reminiscent of their childhood and seemingly light-hearted play is common. Perhaps playing together helps to reduce the aggressiveness which prevents two animals from entering the close contact needed for mating to take place, and it could help to promote the bond between the pair.

There are two kinds of copulation: short (up to two minutes' duration) and long (up to an hour or so). The significance of the times and dates of mating is not altogether clear but long copulations take place only when sows are fully on heat and the short copulations do not result in conception (Neal and Harrison, 1958). As the long copulations, which result in fertilization, are most frequent at the springtime mating peak, this is best regarded as the main mating season. Ovulation is induced by copulation and takes place as early as February but there are further ovulations in June, September and October even in pregnant sows.

The previous litter of cubs may be less than a month old during the spring matings but this does not mean that the sow is going to be burdened with developing embryos while still suckling. The embryos show little sign of development until the following winter, in December or January. Only eight weeks then elapse until birth. The delay in development is due to delayed implantation. The environmental factor which triggers implantation is in dispute, as is the function of delayed implantation. In badgers, it seems to allow the animal to adjust mating and birth seasons to the best advantage, but this cannot be the reason in other animals. Badger cubs are born mainly in February in England, earlier farther south and later in the north. The peak of births is in March in Sweden (Notini, 1948), which gives them ample time to grow to independence by the following autumn. Gestation takes only eight weeks, which, without delayed implantation, would mean that mating would have to take place in December or January. This is a bad time because midwinter is the time of food shortage when badgers are eking out their food reserves in the warmth of the sett, and cannot afford the considerable activity and expenditure of energy of courtship. Therefore there is a decided advantage in the mating season being brought forward to a time when food is plentiful and the badgers are active.

The cubs are tiny at birth, being no more than 15 centimetres from head to tail. They are furred, but the facial stripes are indistinct until a few days old, and their eyes remain shut for ten days. The litter usually consists of two or three cubs with a record of five. For the first six weeks of life they stay underground in a chamber not far from the entrance to the sett. While the cubs are confined underground the sow behaves furtively and does not move far from the sett (Neal and Harrison, 1958). Later she bustles out to find the food needed to supply the milk for the growing cubs and she returns at intervals during the night to suckle. The first steps taken by the cubs above ground are very cautious. The sow appears first and checks that the coast is clear before leading them out. They keep so close as to be hidden by her body and scuttle back to the sett at the slightest hint of danger. Boldness develops gradually and the cubs spend more time above ground and wander farther afield when they are twelve weeks old.

An important part of the cubs' time above ground is spent in play, which becomes quite boisterous, and the parents will join in. Pell-mell chases ending with all-in scrimmages and 'King-of-the-castle' on fallen trees are the main games. Gradually, play gives way to more serious activities. Weaning starts at about thirteen weeks and takes two weeks. The cubs have to learn to forage by following their parents' actions and they also indulge in short bouts of digging and gathering bedding.

By October, the cubs are ready to leave their parents. They are two-thirds grown and weigh 7–9 kilograms. Some do go off to live in separate setts, sometimes sharing with other youngsters, but many cubs spend their first winter with their parents and depart in spring. Sows become mature when a year or so old and males when two years.

During the course of a year the number of badgers living in a sett fluctuates and a particular sett may be abandoned for periods. Dunwell and Killingley (1969) record one family visiting eight setts during the course of a summer. Some setts are large, the homes of one or more families, but others never have more than one or two entrances and give the impression of being of less importance to the badger community. Breeding usually takes place in those setts with several entrances. Occasionally there will be more than one pair of adults occupying the sett. So what are the social relations of the badger?

The interactions and relationships between animals are not easy to study without some means of identification and tracking the movements of individuals. The badger's social life is now being revealed through the use of radio transmitters strapped to the animal's back and by the technique of marking food baits with coloured plastic strips (Kruuk, 1978). The strips are later found in the dung pits. The basic unit of badger society is the sow and her cubs. Two or three sows may live together in one large sett or in neighbouring small ones. They will be accompanied by one or more adult boars.

This clan numbers six to twelve animals (Neal, 1977) and occupies a communal range of about 50 hectares (Kruuk, 1978) containing several setts. The size of the clan fluctuates with the number of cubs and youngsters in residence. Hans Kruuk has found a bachelor clan consisting solely of boars of different ages. Yearling badgers tend to live in outlying small setts with only one or two entrances. The main setts are occupied by the adults and cubs. Whether there is a hierarchy among the boars is not known for certain. Sows have been seen mating with more than one boar in succession, even with strangers from outside. The boars occupy different sleeping chambers from the sows who defend one part of the sett against all comers when the cubs are small.

The home range is defended by the boars fighting and scent-marking, particularly in the early part of the year. Later in the year, boundaries are not so rigid and the badgers wander farther. Marking with musk has already been described and the dung pits are a second form of territorial marker. As well as being located near the sett, they are scattered along the range boundaries, usually near a landmark, and act as massive scent beacons. Phil Drabble describes what appeared to be a violent territorial battle in which he had to rescue his tame badger from being severely mauled by a wild male badger, yet wild boars may live happily together and compete for a sow with little overt aggression (Paget and Middleton, 1974).

A tantalizing glimpse of badger social relations is given by reports of the burial of dead badgers. Vesey-Fitzgerald (1942, and quoted at length in Neal, 1948) describes a sow digging a hole and, with the aid of another badger, dragging the corpse of a boar from the sett and burying it in the hole. Bronwen Doncaster (quoted by Neal, 1977) saw a badger drag a corpse across a road and up a bank. Dean (1949) saw a dead boar pushed out of the sett by the sow so that it rolled down the slope in front of the entrance. Five loads of bedding followed, which the sow scratched over the corpse, and she finished the inhumation with earth, sand and stones. Barlow (1949) was shown the badger equivalent of a prehistoric round barrow. The dead badger was buried under a mound of earth, made by digging out a circular ditch, heaped up and patted down to a plaster finish.

OTTER
(Lutra lutra)

The long lithe body of the otter is streamlined for swimming. The toes, set on short legs, are completely webbed and the head is broad, flattened from above. To see an otter displaying its agility is to see an animal wonderfully at ease in water, yet the European otter is not so well adapted for an aquatic life as a seal or even its near relative the sea otter of the North Pacific coasts. The otter enters water to hunt and to travel but it is not immersed for extended periods. Indeed, if compelled to stay in water for long periods it becomes chilled, so it is best to refer to the otter as semi-aquatic.

In comparison with sea otters, the skeleton of the European otter is not greatly adapted for an aquatic habit. The post-cranial skeleton is not modified (Sokolov and Sokolov, 1970) but there are a few changes in the forelimbs which can be associated with a swimming habit (Ondrias, 1961). The humerus is stout and curved but not so curved as a sea otter's, neither has it the same oval cross-section. On the other hand, the spine on the scapula is well developed, whereas it is reduced in sea otter and harp seal. There are no adaptations in the bones of the feet, although the muscles are modified for regulating tension in the webs. Outside the skeleton, aquatic adaptations are more in evidence. The tail is muscular, flattened, and broad at the base. The nostrils, eyes and ears lie along the top of the head so all three senses can be brought into play when the otter is almost wholly submerged. A similar arrangement is seen on the head of a hippopotamus. The nostrils and ears are closed when the otter submerges. The external ears are small but are sufficiently obvious to be seen when an otter's head pops up.

The fur is waterproof. The dense underfur traps an insulating layer of air and the long guard hairs are oily and water repellent. When an otter dives, the guard hairs lie flat over the underfur to make a streamlined, waterproof covering. When it climbs out, water runs off the guard hairs and they bunch together, gathering at the tips to give the coat a spiked effect. One shake and the coat is almost dry, as water is thrown off the guard hairs. An otter was weighed as it dried itself and it was found that less than 25 grams of water had been left on the fur after the first shake.

In internal anatomy, adaptations for diving are seen in the large volume of the lungs, the right lung having four lobes and the left two lobes. The maximum duration for dives appears to be three or four minutes (Harris, 1968) but they usually last for less than one minute (Hewson, 1973). While underwater, the otter leaves a tell-tale trail of bubbles, partly from air leaving the underfur and partly from air streaming from the sides of the mouth.

The otter is one of the larger mustelids. The head and body length is 62–83 centimetres, the tail being over one-third of the total length, 36·5–55 centimetres. There are records of dog otters measuring 15·2 centimetres. Average weight is 6–15 kilograms with records of nearly 27·2 kilograms. Female otters are about 15 centimetres shorter and 3 kilograms lighter than males.

The underfur is whitish-grey with brown tips but the overall colour of the otter is given by the glossy guard hairs which are grey with rich brown tips. The colour of the back is a dark glossy brown, often darker on the head. The flanks are rather lighter and the underparts are

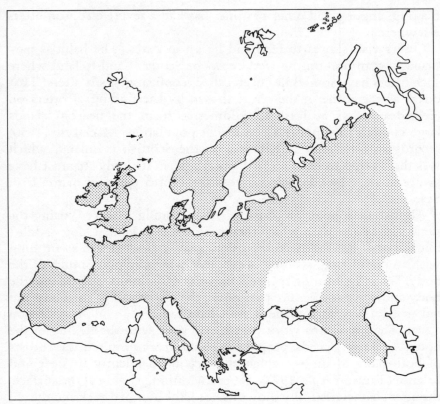

greyish, the pale fur reaching around the neck to the ears. White or white-spotted otters have been recorded (Harris, C. J., 1968).

The European otter ranges throughout the continent from the Arctic Circle southwards. It also occurs in North Africa and across Asia to Japan. Its habitat is aquatic, both freshwater and marine. Otters are largely limited to watercourses and lakes but they sometimes wander some distance overland, particularly when crossing a watershed from one river system to another. They prefer fertile, eutrophic waters including marshes and saltmarshes where there is plentiful food. Stephens (1957) considers small rivers running directly into the sea as the best otter habitat in Britain. Bankside vegetation is important, either as cover for the otter or as a breeding ground for prey, and clearing river banks has an adverse affect on otter populations. However, otters are regularly found in barren, stony-bottomed highland streams and lakes but such places are little disturbed and a combination of bank clearance and increased disturbance in the form of boating, angling and other pastimes has had a severe effect on otters elsewhere.

Otters visit shores to feed and lie up in caves. The habit is particularly common on the west coasts of Scotland and Ireland where such otters have sometimes been called, confusingly, sea otters. That otters are at home in the sea is shown by the finding of otters on Hebridean islets at least 13 kilometres from the nearest islands large enough to support a permanent population. Macintyre (1950) records an otter shot in Kintyre, on the Scottish mainland, which was discovered to have an Irish spearhead in its body. It must have swum at least 24 kilometres across the North Channel of the Irish Sea.

Almost all animals are capable of swimming if forced, using the same sort of gait with which they progress over land, a 'dog paddle', but the otter has followed the fishes, seals and whales in swimming with an undulating, sculling movement which involves much of the body. When swimming at speed, the otter's forelimbs are held into the body and waves of flexure pass down the lower half of the body and tail, in a dorso–ventral plane as in a whale, rather than side–to–side as in a seal. The hindlimbs work together in time with the flexing of the body. At slower speeds, propulsion comes from simultaneous thrusts of both pairs of limbs, which also act independently to steer and balance (Tarasoff et al., 1972). When swimming slowly at the surface, the otter characteristically shows its head, the humped back and the tail lying flat on the surface.

On land, the otter is agile despite its short legs. It walks with the head held low and outstretched. The body is inclined at a slight angle because the hindlegs are slightly longer than the forelimbs. As the otter moves faster, the back becomes arched and the tail is held off the ground. Top speed, as fast as a man can run, is achieved by bounding with the two pairs of legs working together, but with the right foot of each trailing slightly. When travelling on snow or ice, the otter economizes by alternating running with sliding. The hindlegs give a push off and momentum is maintained on flat surfaces by a sideways flexing of the body.

As has been mentioned, an otter's main sense organs lie on top of the flat head where they can be used while the animal is swimming. The nostrils and ears are closed while submerged. The use of the sense of smell underwater is ruled out but the otter can still hear by sound being conducted through the bone of the skull, although it will not be able to locate sounds. An animal pinpoints the source of a sound by noting the difference in arrival of the sound at each ear. Sounds cannot pass easily from air into bone so the skull forms a barrier around which the sound has to pass into the ears. However, sound waves readily travel from water into bone so the outer ears are by-passed and sound location is impossible underwater. Whales avoid this problem by their ears being insulated from the skull and sounds have still to enter through the usual channels, but there is no such adaptation in the otter.

The eyes are small but sensitive. No studies have been made on the vision of the European otter but the visual acuity of the Asian clawless otter (*Amblonyx cineria*) is the same when submerged as in air, and is about the same as for other mustelids (Balliet and Schusterman, 1971). The upturned angle of the eyes could make it easy for an otter to spot fish and waterbirds by swimming under them and spotting their silhouettes. Such a trick has the added benefit of enabling them to avoid detection by prey who cannot easily look down. With the nostrils and ears closed, the eyes must be important when hunting in water but they are rendered useless in murky water or at night. In such situations it seems that the otter must make use of its magnificent array of long, sensitive whiskers. These are set in fleshy pads on each side of the snout, with one or two on the elbows. The whiskers must be excellent for feeling objects in the dark or when grubbing in the mud of lake bottoms but Harrison Matthews (1952) has suggested that whiskers could also be used for detecting distant objects by recording turbulence in the water.

Numerous studies have been made of the diet of otters by exploiting the ease with which faeces can be found. Known as 'spraints', faeces are used as scent marks and are deposited in conspicuous places (see below). The most useful analyses are those in which there has been a parallel investigation of what is available to the otter. Marie Stephens (1957) examined otter corpses and spraint collections from all over Britain and reviewed past work. Fish, expectedly, make up the bulk of the otter's diet but, in general, the emphatic conclusion from several studies is that otters are not serious pests of salmon and trout as has often been claimed. In Poland, otters were exterminated and fish began to decrease as disease struck them hard—the otters had been pruning the population of sickly fish. Like so many predators they are economical of their time and energy and prefer to chase easily caught prey. The strength of the salmon family, which makes them so prized by sportsmen, renders them less easy for otters to catch. Nevertheless, otters living in a stream containing little other than trout can hunt little else. But it is quite likely that the otters catch the less fit fishes, thereby weeding out those suffering from disease or injury and so helping to maintain a healthy population.

Erlinge (1967) has made a detailed study of otter feeding habits in the lakes and streams of southern Sweden and found otters had no effect on fish populations. Crayfish, a favoured food, were subjected to a heavy predation but they maintained their numbers. Only in fish hatcheries do otters become serious pests, as they may indulge in an overkill, analogous to the fox and the hencoop, in the presence of a superabundance of helpless prey.

Erlinge's study showed that if otters eat a high proportion of one species, it is because that animal is easily caught, which means that the behaviour of the fish species determines whether it is caught. Captive otters select the easiest prey to catch if they are hungry (Erlinge, 1968b) but when well fed they can afford to pick and choose. In this way Erlinge has demonstrated that the trout's speed is the reason for the small number of them caught. If trout are slowed down by snipping pieces off the tailfin, even well-fed otters catch and eat them as an extra snack, in preference to other species. Trout and salmon, then, are a delicacy that is hard to obtain, rather than a dull food to be eaten when nothing else is available. It is often pointed out that otters have a preference for eels over gamefish, but this is most likely to be because they are slow-moving and easier to catch.

Erlinge's otters had a seasonal variation in diet as a particular prey species became available. He found that three-quarters of their food was fish, made up of cyprinids (mainly roach and bream), percids (mainly perch), pike, eels and sticklebacks. Pike are eaten in the spring spawning period when they gather in shoals, and eels are eaten during their active period from May to October. Small fish are eaten when they gather in shallows in spring and autumn. Tench were abundant in the lakes but were rarely eaten because they keep to dense cover. Fewer fish are eaten in summer months, perhaps because they are more active and harder to catch, and their place is taken by birds or crayfish. Young birds are easy prey at this time (see below) and, in contrast to fish, the crayfishes' activity in summer makes them easier to catch. Crayfish are rarely found in spraints during the winter because during that period of the year they retire to crevices and under stones.

The flexibility of the otter's diet was demonstrated when the summer crayfish suddenly became scarce after disease struck in Swedish rivers. The otters then turned to frogs and fish, which normally are not much eaten during the summer. Frogs are sought mainly in spring when they gather to spawn or in winter because, unlike crayfish, frogs hibernate on the open bed of pools. Hibernating fish are also sought, as in the Kerzhenets River in the USSR where fishermen approve of otters because they disturb the fish in the pools where they lie in winter so that they swim downstream to the waiting nets (Ognev, 1931).

In the different habitat of saltmarsh and river on the Norfolk coast, Weir and Banister (1971) found the same pattern of predation. The very abundant three-spined sticklebacks were taken throughout the year, the large shoals being very vulnerable, while eels and crayfish were popular during the summer months. Cyprinid fishes were taken mostly outside the summer. Trout were fairly common in the rivers but bones were found in spraints on only one occasion. Surprisingly, only a few marine fish appeared to be eaten in this habitat.

Other aquatic animals make up a small part of the otter's diet. These include dragonfly nymphs, water beetles, water snails, freshwater mussels and crustaceans such as the freshwater shrimp *Gammarus*. While frogs are commonly eaten, with the reproductive tracts of the females and their huge 'jelly glands' being rejected, toads are rarely eaten, perhaps because of the poison glands in the skin. Along the shore, conger eels, lobsters, various crabs (*Carcinus, Cancer, Portunus*), fishes (flatfish, wrasse and cod) and small crustaceans (*Porcellio, Ligia, Orchestia, Idotea* and *Gammarus*) have been recorded as otter prey.

Of warm-blooded prey, a certain number of birds and mammals are taken. As with fishes, the criterion for inclusion in the diet is ready availability. Inexperienced or flightless young birds are the most frequent victims. The list of birds includes ducks, coots, moorhens, which are taken occasionally, and heron, grebe, lapwing, golden plover and starling, which are eaten very rarely. An otter was once seen bringing a gannet ashore (Macintyre, 1950) and another caught a black-headed gull on a lake, despite the attacks of other gulls (Cozens, 1959). Rabbits are the most frequent mammalian prey, according to Erlinge, followed by water voles, field voles and moles. The occasional hare and shrew can be added to the list. Mammals may be more important in severe winters when freshwater prey are imprisoned under thick ice, and birds, mammals and fish are always welcomed as carrion.

Compared with the volume of data published on faecal and stomach analyses, there have been very few observations on the feeding behaviour of wild otters. A Canadian otter was caught in a crab pot at 20 metres but most hunting is carried out at shallower depths. Hewson (1973) has watched otters feeding in a Scottish loch. Dives lasted about fifteen seconds and the time spent on the surface between dives averaged six seconds. This is little more than is needed to exhale and take a new breath. While capable of diving without a sound when disturbed, an otter will make quite a splash when feeding. It arches its back and 'duck-dives' with its tail held in line with the body and whipping vertical when it sinks, as do a human diver's legs.

In Scotland, but not apparently in England, food is brought ashore to a regular 'table' which can be recognized by the remains scattered about. Salmon may be abandoned with only a lump taken from the body, to the dismay of the angler and delight of the casual diner. At first sight this is wanton destruction but the lump represents a square meal for an otter. Otters are rather fastidious feeders, and instead of bolting its food and breaking up the carcase like a fox, an otter eats deliberately. The front teeth are sharp and suitable for seizing slippery prey but the molars are flattened, rather like a badger's. Food is held down with the paws and morsels are pulled off and well chewed before being swallowed. When eating a bird, the otter leaves the flight feathers and larger bones intact.

Where they are undisturbed, otters are active by day but human activity restricts their activity to night and the twilight hours. The result is the otter's acquisition of a reputation for secretive habits, as exemplified by the Otter in *The Wind in the Willows*. Yet in its natural state the otter appears to be no more retiring than other carnivores. It is instructive to read Dugald Macintyre's *Wild Life of the Highlands*. Macintyre spent his life on the moors and shores of Argyll and gives many accounts of finding otters hunting or playing in broad daylight without any trace of shyness. Alas, for the otters' confidence, the anecdote usually ended with their death by trap or gunshot! Macintyre saw so much of otters because his working hours were spent in otter habitat and so he was likely to come across them. To search for otters will probably be unrewarding unless many patient hours are spent. As Marie Stephens says, by far the best way to see otters is to take up fishing. Chance encounter can be rewarding. An otter may swim past as one sits quietly by the shore or, rounding a promontory in a boat, one may disturb an otter as it feeds on a rock or writhes on its back, little legs in the air, enjoying a good scratch.

The otter has often been described as a nomad, spending a short time on a stretch of water then moving on and having no permanent habitation. Signs that an area is occupied by an otter, such as spraints and food remains, tend to confirm that otters occupy then abandon a particular spot but detailed study of such signs by Sam Erlinge (1968a) in southern Sweden has revealed the pattern of the otter's life, although it must be kept in mind that not all European otters necessarily behave in the same way as those that Erlinge studied in a complex of Swedish lakes and streams.

Erlinge tracked the otters by following footprints on sandy shores and over snow in winter but there were several other signs of otter activity. By travelling in water the otter leaves little sign of its movement, but when the water is frozen it follows the same course over the ice. At intervals, however, it comes ashore and its regular passage is marked by well-worn paths up banks, around waterfalls and as short cuts across headlands. Some paths continue to be used for decades. When the ground is snow-covered, or on suitably slippery conditions at other times, the otter resorts to sliding as an economical or playful form of travel, but Erlinge did not find any special sliding places. If there are favoured hunting grounds, a nearby feeding place reveals what is being caught and small patches of bare earth are rolling places. Otters delight in rubbing their backs and rolling about as part of the grooming and drying process.

An otter has several places where it lies up to sleep. Some are casual resting places under boulders and piles of brushwood or in cavities under the exposed and overhanging roots of trees, in drains, rabbit holes and fox earths, even in pollarded willows. Caves are used around seashores. The otter may sleep on bare earth or on a couch of vegetation. Couches are made above ground on sheltered islands or in reed beds. They consist of piles of stems or twigs gathered from nearby and laid parallel. Couches described by Hewson (1969) on a Scottish loch were oval or nearly rectangular and measured 46 by 30 centimetres to 91 by 76 centimetres. The centre becomes depressed through use so that an old couch resembles a large bird's nest. One of Hewson's couches was in use (with repairs and additions) nine years after he had first found it.

The underground den or holt is an elaborate affair where an otter has enlarged a cavity or burrow for its own use. There may be more than one entrance to the sleeping chamber, including one opening underwater but this will depend on whether the soil is workable. Underwater entrances are impossible where bedrock lies near the surface. As with a badger's sett, there may be a side chamber acting as a latrine.

Current use of any of these resting places is immediately indicated by the presence of spraints. As it goes about its daily round the otter deposits small piles of droppings tainted with secretions from the anal glands to act as scent-marks, equivalent to the urine of dogs and the dung pits of badgers. Their role will be described later and their use in determining otters' diet has already been noted. Spraints are deposited by sleeping places, rolling places and runways and otters also leave the water to deposit spraints along their line of travel. They choose conspicuous features and it is easy to learn what is likely to be conspicuous to an otter with its eyes and nose just clear of the water. Free-standing boulders, spits formed by the confluence of streams, under bridges, and the angles formed where a stream runs in or out of a lake are favoured sites. Spraint is often placed on large, conical boulders over a metre high, indicating considerable agility on the part of the otter. Sometimes a U-shaped runway indicates where otters climb out of the water, set spraint and return. The otter, itself, sometimes makes the spraint more conspicuous by placing it on a small pile of sand or mud, or on a twist of grass. Instead of spraint, scent from the anal glands alone may be deposited on these little hillocks. When fresh a spraint is black and glistening with mucus which, it has been suggested, may ease the passage of fish bones through the intestine. The spraint is packed with fish bones when fish has been the diet of the otter, and there is a not unpleasant fishy smell. As the spraint ages, it dries up, becoming grey and powdery.

From all these clues of activity Erlinge has been able to piece together the otter's private life. Each individual occupies a length of waterway where it has its places to sleep and roll and its favourite routes which are studded with sprainting points. A dog otter occupies a stretch of water about 14 kilometres long. This is its home range where it spends its adult life and the false impression of otters as nomads is partly explained by an individual abandoning parts of its range for a period. There may be, however, general movements, for instance, to the spawning pools of frogs.

An otter will travel many kilometres in a night but it tends to relate its activities to the prevailing circumstances. A small part of the range may be occupied for some time before the otter moves on. In winter, ice may restrict its movements to streams which remain free of ice, whereas the spring thaw encourages the otter to make a thorough investigation of the range. The summer sees a more sedentary life and autumn brings more activity associated with changes in diet and sexual activities.

Otters are basically solitary when not involved with mating or rearing families. Their lives are so arranged as to keep apart, yet they must also keep in touch. Communication is effected through the medium of the scent-marks which must communicate to other otters the depositor's identity, sexual status and, perhaps, give an indication of the time which has elapsed since it passed by. The value of scent-marks

in communication has led some investigators to abandon the collection of spraints for analysis out of a fear that the social life of the
otters may be disrupted. Imagine studying human society by stealing
letters before they were delivered!

The territorial behaviour of otters follows the pattern of the weasel
and stoat (Erlinge, 1968a). The dog otter occupies a range which is not
quite a territory because its boundaries overlap the ranges of
neighbours. The only portion likely to be seriously defended is the
central area where the otter spends most of its time. Claim to the
remoter parts of the range is staked by scent-marks and the resident
makes special journeys to 'beat the bounds' by renewing scent-marks.
Where the population is dense (Erlinge recorded one otter per
kilometre along one stream in winter), there is very intense marking
as ranges overlap considerably, but when otters are well spaced marking sites go out of use.

Depositing scent and visiting other otters' marks keep the otters
informed about their neighbours. If one dies, the vacuum is soon filled
by neighbouring otters expanding their ranges or by a newcomer
settling in. Actual attacks or pursuits are rare because the otters take
care to keep out of each other's way. Conflict is most likely to occur
when a newcomer arrives. These animals are wanderers like the transients in the weasel populations. Sometimes they can settle for a while
on the edges of established territories and live as subordinate animals.
They avoid the residents by noting and avoiding marking places and
travelling by water.

Female otters live with their cubs in smaller ranges set within those
of the dog otters. One dog may have two or more bitch ranges wholly
within or overlapping his range. The dog is dominant and may cause
the bitch to shift her range. For preference she chooses an area where
food is plentiful and, as the cubs grow, her range extends to provide
the necessary extra food. Because the ranges of the family parties are
concentrated in optimum habitats, intervening poorer ground is left
untenanted by bitches. Their ranges do not often meet or overlap,
except when the cubs are well grown, but even then families seem to
avoid each other so there is little sign of aggression between females.

One of the reasons for a dog otter's patrols of marking places is to
discover whether or not the bitches living around his range are in
oestrus. Otters in Britain give birth in any month of the year, in so far
as this is known, whereas in Sweden births are in the summer and
mating in spring. Whether this has any connection with the harsh
Swedish winter is not known. Such a condition could be attributed to

the difficulty of finding food during the winter freeze but very little is known about the reproductive physiology of otters. Delayed implantation occurs in the Canadian otter but not, apparently, in the European species.

Reproductive behaviour is also something of a mystery and most of our knowledge comes from captive otters, and then mainly the American otter. Dog and bitch associate for several days and Liers (1951) suggests that the bitch is on heat for 42–46 days with periods of receptivity six days apart. Mating takes place at short intervals in the water. The act may last, apparently, for as long as an hour (Cocks, 1881) but usually for much less and the bitch whistles or caterwauls throughout.

The cubs are born with a silky, pale-grey coat some nine weeks later, and their eyes open when five weeks old, the normal litter size being two or three, exceptionally up to five. The litter is born in a couch or a holt, in a quiet stream or backwater or in a remote pool, and the bitch becomes very intolerant of the dog's proximity at this time (at least in captivity and it is fair to conclude that the same applies to the wild). There are records of the dog being seen with the family but this occurs when the cubs are several months old and is probably a rare occurrence.

The bitch is an exemplary mother. For the first seven weeks the female licks up the cubs' faeces as they are voided (Anon., 1971), after which they become house-trained. Harris's captive Canadian otter hauled her seven-week-old cubs in and out of the nesting-box to house-train them (Harris, C. J., 1968). The cubs take to the water for the first time when seven or eight weeks old. They are wary of leaving *terra firma* and have to be coaxed by their mother. Harris's otter entered the water and faced the cubs, calling. Failing to overcome their reluctance, she grabbed each one by the scruff of the neck and pulled it in. Once in, the cubs are quite happy with their new medium although a little practice is needed to gain full control. The family goes on forays together, mother leading and the cubs in V-formation or a straight line astern. The mother has been seen to come to the rescue if the cubs get caught in rapids and she helps them learn to feed themselves (Stephens, 1957). The bitch has been seen to bite an eel in two for the cubs and, if necessary, chew it into small pieces for them. The bitch is very aggressive in the defence of her cubs and there are several anecdotes of otters attacking dogs and men. The cubs stay with the parent throughout their first winter and into the next spring so that bitches probably do not breed every year.

The Mongoose Family
(Viverridae)

The viverrids are a varied group of small carnivores whose most familiar members are the mongooses and civets. In many respects the viverrids resemble the mustelids, particularly in their short legs and long body, and in the possession of the anal glands. Secretions from the civets are used as a base for perfumes. In general, viverrids can be distinguished by their tapering heads which have little or no break between muzzle and cranium. Many species have striped or spotted coats.

Viverrids are mainly nocturnal and many are arboreal. They are plantigrade or digitigrade, the latter group having semi-retractile claws. The teeth are less well adapted for meat eating than in the mustelids, with the carnassials being poorly developed, and some viverrids subsist largely on fruit, insects and other small animals. Some mongooses are social and live in bands. On the whole, the biology of the family is not well known.

The range extends over the warmer parts of Africa, Asia and Madagascar. The feline genet and Egyptian mongoose have been introduced to southern parts of Europe. The seventy-two species are grouped into six subfamilies. The Viverrinae includes the civets, genets and linsangs; the Paradoxurinae, the palm civets; the Hemigalinae, the banded civets; and the Herpestinae, the mongooses. The remaining two subfamilies are confined to Madagascar: the Galidiinae, the Madagascar 'mongooses', and the Cryptoproctinae, the fossa.

FELINE GENET
(Genetta genetta)

The genets can be described as a mixture of cat and weasel. They belong to the same family as the mongooses but they have the alert poise and graceful movements of a cat and sinuous shape of a weasel. The six species are distributed over Africa, with the feline genet occupying most of the continent outside the desert areas, also Palestine and Arabia, and penetrating into south-west Europe. There are records of genets being sold in European markets as pets. They tame well if caught young, and they are as attractive as any cat and are good mousers. However, they possess anal glands secreting unpleasant odours, though surely no worse than any tom-cat.

The feline genet is also known as the small spotted genet. This is an aptly descriptive name as the fur of the body is spotted and streaked with brown or black on a silver-grey background. The head is pale, the dark cheeks contrasting with whitish colouring under the eyes and on the tip of the muzzle. The tail is ringed with light and dark bands. The head has a pointed muzzle, more fox-like than cat-like, and the sharp ears are mobile in the extreme. The paws are delicate and the claws retractile. The head and body is 47–58 centimetres and the tail is almost as long again at 41–48 centimetres. Height at the shoulder is 18–20 centimetres and the weight is 1–2·2 kilograms.

The European range of the genet is Spain, Portugal, the Balearic Islands (Corbet, 1966) but it is now rare in France. Its habitat is thick, dark woods where there are hollow trees and rock crevices for dens, although genets will also use the abandoned holes of other animals.

Very little is known of the habits of genets in the wild. They are nocturnal, solitary and elusive, and what little is known of them has come from observations of captives. One is struck at once by their agility and grace in progressing along branches. The sharp, retractile claws give a good grip and the long tail acts as a balancer. To operate efficiently at night, it is likely that a genet learns intimately the details of its home area. A captive genet put in a new cage explores its surroundings with intense scrutiny. It investigates everything around it with nostrils working, ears twitching, long whiskers probing and eyes glaring. Each step is tested until the foothold is assuredly secure. Having made one circuit, the genet goes round a second and third time

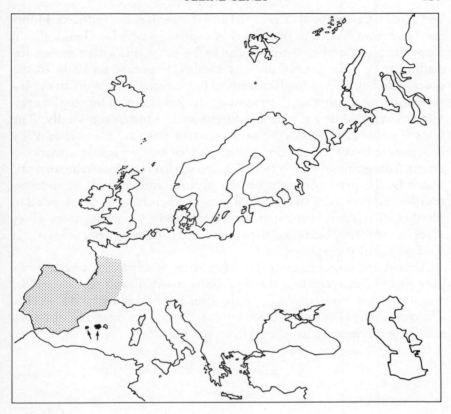

and from then on it can race about its cage in total darkness without putting a foot wrong. The whole of its surroundings must have been memorized (Burton, 1955).

That genets are solitary is indicated by captive individuals having nothing to do with each other even when confined to the same cage. The father will even kill his own offspring if left with them (Gangloff and Ropartz, 1972). It would seem likely, then, that genets maintain defended territories or, at least, keep well out of one another's way. Scent-marking with the anal glands is accomplished in a handstand position, the genet standing on its forepaws to wipe the glands against a vertical surface (Ewer, 1968) and by rubbing the flanks against objects. When alarmed by a human intruder or on meeting another genet in an aggressive situation, the genet erects a mane of black hair along the back and fluffs its tail into a 'bottlebrush'. At the same time it hisses like a snake and bares its teeth.

Genets are very much hunters rather than omnivores. The canines and carnassials are well developed and the molars are reduced. Hunting behaviour has been described in captive genets by Gangloff and Ropartz. Sight and hearing appear to be most important senses for finding prey. The genet hunts by stealth, lowering its body to the ground and sliding along like a snake before attacking with an explosive rush. Small animals, such as mice, are killed with a bite to the neck but rats are held down with the forepaws and bitten repeatedly. The bites are not orientated to the neck, as in the polecat for instance. Prey is consumed entirely, starting at the head, or rarely a leg, and working down. Sometimes the prey is held down so that chunks can be torn off. Much of the prey is caught on the ground and consists of rodents, reptiles and insects. Scorpions and grasshoppers have been recorded in the diet of African feline genets (Ewer, 1968). Climbing trees gives access to roosting birds and their nests and captive genets, at least, eat fruit (especially grapes).

Almost nothing is known about breeding. In captivity, two litters a year have been recorded, the first born in April and the second in August or September (Volf, 1968), after gestations of 10–11 weeks. The usual litter is of one to three kittens. Their eyes open at eight days and they are weaned at twenty-five weeks.

EGYPTIAN MONGOOSE
(*Herpestes ichneumon*)

The Egyptian mongoose is found throughout Africa, except for desert and dense forest regions, as well as in Palestine, but its European distribution is largely confined to the southern end of the Iberian peninsula. It seems likely that its presence in Europe is due to human agency. In Ancient Egypt, the mongoose was sacred to Mafdet, a goddess who gave protection against snakebite and it was once fashionable for Roman ladies to keep pet mongooses (Zeuner, 1963). The introduction of mongooses into Iberia could have been either as pets, as cult animals which later went wild, or as a means of controlling snakes. Mongooses were introduced to Italy for this purpose in the 1960s (Hinton and Dunn, 1967) despite experience in the West Indies

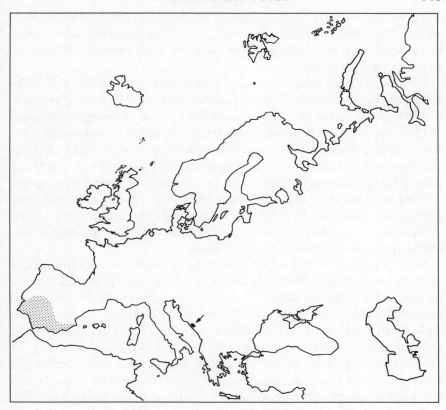

and elsewhere that the only result of such introductions is the establishment of a new pest.

The Egytian mongoose has the alternative name of ichneumon, from the Ancient Greek for tracker or pursuer. It is distinguished from similarly short-legged mustelids by the pointed snout which runs evenly into the head, and the tail which tapers from a stout root: two characteristics which combine to give mongooses a very streamlined appearance. The rough coat is greyish-yellow, darker on the back, and the tail ends in a tuft of black hairs. The ears are short and broad, with the appearance of being set low on the head. The claws are long and sharp but not retractile. The head and body length is 51–55 centimetres, tail 33–45 centimetres, shoulder height 19–21 centimetres and weight is 7–8 kilograms. The sexes are similar.

At one time mongooses were widespread in Iberia but they are now restricted to the south, around the Sierra Morena, and are generally rare or locally abundant within this region. As well as the recent introduction to Italy, mongooses have also been established in Dalmatia. The preferred habitat is scrub, maquis, woodlands where there is an abundant shrub layer, and reed- or rush-covered fringes to marshes and swamps.

Compared with other species of mongoose, little is known of the habits of the ichneumon. It has been barely studied in Africa, where it is widespread, but some surveys have been made of mongooses living in the Coto Doñana Reserve. They are not so strictly nocturnal as was once supposed and may be more active during the day, particularly in the morning before the sun becomes too hot. They hunt singly, in pairs or in family parties and must presumably occupy home ranges similar to those established by mustelids. The closely related Indian mongoose scent-marks with its anal glands by cocking its leg or performing a handstand to set the mark as high as possible. It also marks with facial glands (Ewer, 1968).

Surprisingly little is known about the breeding of any mongoose species. From observations of captive ichneumons, mating takes place in April and gestation lasts 104 days or less (Ewer, 1973). Mating is preceded by a chase in which the female runs from the male who squeaks in the typical social-contact call. The female repeats the call, then stops and crouches. As soon as the male catches up, she sets off again. This coquettish behaviour is repeated several times before the female assumes the mating position with tail twisted to one side and the male is permitted to mount. The litter consists of two to four kittens. Nothing is known of their development although, by analogy with other species, they are probably weaned at one month. After they have left the nest, the kittens follow their mother in single file, each animal tucked under the tail of the one in front.

Until the outbreak of myxomatosis, Iberian mongooses fed mainly on rabbits. Even where these are still abundant, other vertebrate prey such as rodents, birds, reptiles and amphibians, as well as insects, are included in the diet. Mountfort (1958) describes mongooses of the Coto Doñana entering the nest-burrows of bee-eaters. Mongooses are famous for two specialized methods of obtaining food: catching snakes and opening eggs. Spanish mongooses catch poisonous Montpellier snakes and owe their success to the lightning speed of the strike. The reptile is seized and fatally bitten on the neck before it can strike back. A small egg is held in the paws and a hole is bitten in the narrow end but larger eggs are held in the forepaws and hurled backwards between mongoose's legs against a rock.

The Cat Family
(Felidae)

Second only to the dogs in familiarity, the cats have not been bred for such a variety of purposes as the dogs. The domestic cat is widely used as a laboratory animal but otherwise its domestic uses are restricted to providing company and controlling rodents. Compared with the dog family, the cats have a short muzzle and short braincase, with broad zygomatic arches, giving a rounded head. The eyes are set in the front of the head to give good binocular vision. The fur is soft, often marked with stripes or spots and is prized by furriers. The feet are digitigrade, five toes (including dew claw) on the front feet and four on the rear. The claws can be retracted, except in the cheetah.

Cats are medium to large carnivores. Their diets are almost exclusively of flesh and they are the most specialized of hunters. Prey is caught after a preliminary stalk or ambush, followed by a short dash. Prolonged fast running is not possible. The prey is despatched by a well aimed 'neck bite'. The canines are slender and sharp, the carnassials are well-developed but the cheekteeth are reduced in number. The tongue is covered with backward projecting papillae and acts as a rasp in feeding.

The cat family is a very homogeneous group. Most of the thirty-seven species are given the generic name of *Felis*. The five 'big cats'—lion, tiger, leopard, jaguar and snow leopard are grouped in the genus *Panthera* and the cheetah is given its own genus *Acinonyx*.

All are solitary, except for the lion, and most are nocturnal. The family is cosmopolitan, except for Australasia. The European species are the lynx and the wild cat, and domestic cats frequently go wild. The lion became extinct in Europe in historical times.

143

WILD CAT
(Felis sylvestris)

Thomas Pennant, the eighteenth-century naturalist, described the wild cat as 'the British tiger'. It is, he said, 'the fiercest and most destructive beast we have; making a dreadful havoc among our poultry, lambs and kids'. And wild cats not only attacked livestock; another eighteenth-century naturalist, the Rev. William Bingley, told of a memorial in a church which commemorated a fight to the death between a man and a wild cat. The fight was said to have started in an adjacent wood and the cat had apparently attacked the man who, fearing that the devil was involved in the assault, had sought sanctuary. He had struggled to the church porch where both contestants were found dead from their wounds. A reputation for ferocity is enhanced by the frequency with which photographs and drawings show wild cats with ears flattened and teeth bared but the reputation is being refuted now that wild cats are being kept in wildlife parks. They seem to be no fiercer than the other inmates.

The European wild cat looks very like the domesticated tabby, but the latter, as far as is known, has been derived from the African cafer or bush cat. The latter probably belongs to the same species as the European wild cat, and the domestic cat will interbreed with both. The main differences between the European wild cat and the tabby are the heavy build of the former and its shortish bushy tail. The skull is robust and squarish and the legs are relatively long. The wild cat stands 35–40 centimetres at the shoulder, has a body length of 47·5–80 centimetres, with a tail of 26–37 centimetres. Females are smaller, and about four-fifths the weight of males. The record weight is of 15 kilograms for a cat shot in the Carpathians.

The fur is long, soft and dense. It is mainly yellowish-grey and there is considerable variation in the markings of dark brown and black. Some cats have vertical bars on the flanks; in others the bars are broken with spots. The face bears prominent dark stripes running from the eyes and the bridge of the nose. The tail is ringed and ends in a black tip. Part of the variation must be due to the interbreeding with the domestic cat, which may occur to such an extent that fears have been expressed that the wild cat as a pure species may become extinct.

At one time the wild cat was distributed widely over Europe, through Asia Minor to the Caspian Sea. It is now found mainly in the

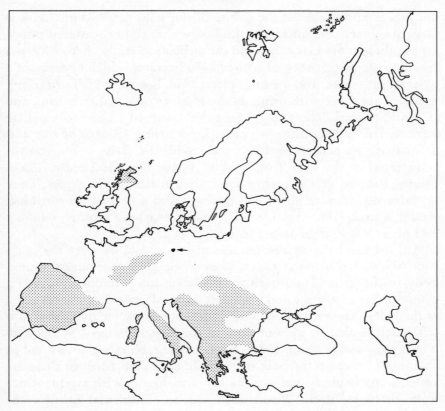

Iberian Peninsula, Italy and south-east Europe, and in the mountainous regions of France and Germany. There is an isolated population in the Highlands of Scotland. The wild cat of Sardinia is thought
to be more closely related to the cafer cat (van den Brink, 1967).

Wild cats have been regularly persecuted as vermin or for their fur.
In 1205 King John gave Gerard Camville permission to hunt wild cats
in the royal forests and throughout mediaeval times wild cat was one
of the few furs allowed to the common people. The flesh was used to
cure gout and the fat to ease pain and dissolve tumours (Hamilton,
1896). Persecution had little effect on numbers and distribution compared with the wholesale destruction by gamekeepers in the
eighteenth and nineteenth centuries. The Rev. Bingley, writing in
1809, described wild cats as being more common than foxes in the
Scottish county of Argyll. Keepering maintained the population at a
low level and with the decline of the big shooting estates since World
War II, the Scottish wild cat has increased in both numbers and range.

Little is known about the habits of the wild cat, except that it is largely solitary and nocturnal. Linked with a thinly scattered population, these traits make the wild cat difficult to study. Each cat has a home range or territory of some 60–70 hectares which consists of a system of tracks and resting places (de Leeuw, 1957). Marking involves spraying with urine, in the manner of a domestic tom, and depositing faeces. These are buried in the centre of the territory but are left scattered around the periphery (Lindemann, 1955). The cats also scent-mark by scratching tree trunks with the claws when scent is transferred to the bark from glands on the feet (de Leeuw, 1957). During the day, cats lie up in a den among rocks or under trees. They occasionally wander abroad in daylight and will bask in sunshine. Adam Watson (1961) found a wild cat asleep in long heather, where it had bitten off heather stems to make a bed.

At least until the advent of myxomatosis, rabbits were the main food of wild cats in many places. They are caught by the same slinking, belly-to-the-ground approach and final dash and pounce employed by domestic cats. Prey is eaten head first, unless large when the cat starts at the belly. The remains of a cat's meal are characteristic: the skin is left neatly rolled inside out around the largest bones (Nethersole-Thompson and Watson, 1974). Even where rabbits are not available, mammals make up the bulk of the wild cat's diet. Most of these are rodents, particularly voles. One cat was found with twenty-three voles in its stomach. Wood mice and, in continental Europe, the closely related yellow-necked mouse, a few rats, shrews and hares are also taken (Condé et al., 1972). Frogs, lizards, and fish are readily eaten if available. Roe deer fawns are occasionally taken and carrion or human refuse dumps are not spurned. Birds seem not to be greatly favoured, and experience with captive cats is that some individuals will eat birds after plucking their feathers, while other cats ignore the feathers.

The breeding season starts in January and is marked by the caterwauling of the tom-cats. Charles St. John, writing in 1845, says: 'I have heard their wild and unearthly cry echo far into the quiet night as they answer and call to each other. I do not know a more harsh and unpleasant cry than that of the wild cat.' The female cat gives birth in her den after a gestation of 63–69 days. The litter consists of one to eight kittens, with fine sparse fur, pale with tabby markings, and whose eyes open after one or two weeks. When suckling, the mother sits back as if reclining in an armchair and the kittens sit on her 'lap' (Meyer-Holzapfel, 1968). The first solid food is taken when the kittens are one month old and they are weaned at three months. Shortly afterwards their mother drives them from the family den (Volf, 1968). There are statements that wild cats have two, and rarely three, litters a year but Meyer-Holzapfel found a second litter exceptional in captive wild cats and suggests that it is unlikely in the wild.

LYNX
(Lynx lynx)

Lynx-eyed is the epithet for extra keen eyesight and the name comes from Lyncaeus, the sharp-eyed pilot of Jason's ship, the Argo. The lynx is built like a large, long-legged tabby cat but has several unmistakable features. The tail is short, the ears are prolonged by hairy tufts and two cheek ruffs combine to give the lynx a distinctive and odd appearance face-on.

There is some dispute over whether Europe has one or two species of lynx, and whether the Canadian lynx is of the same species. According to van den Brink (1970) there are two species: the boreal lynx (*Lynx lynx*) of northern Europe and the pardel lynx (*L. pardinus*) of southern Europe. The boreal lynx belongs to the same species as the Canadian lynx of North America. Its coat is sandy brown with spots on the legs and lower parts of the body. The tail is tipped with black. The body length is 80–130 centimetres with a tail of 11–24·5 centimetres and a weight of 18–38 kilograms. It stands 60–75 centimetres at the shoulder. The pardel lynx is slightly smaller and much more heavily marked. The body, legs and tail are closely set with black spots and streaks. Van den Brink considers the pardel lynx to be ecologically similar to the bobcat (*Lynx rufus*) of southern North America and the caracal (*Lynx caracal*) of North Africa and Asia. The division into two species is not accepted by all authorities (e.g. Corbet, 1966) and there is an alternative suggestion that the two forms represent the ends of a once continuous cline with the Carpathian lynxes as an intermediate.

The distribution of the pardel lynx in about A.D. 1500, before its decline, is given by van den Brink as covering the Iberian Peninsula, southern France, Italy, through the Balkans and Asia Minor to the Caucasus and around the southern end of Lake Baikal. At the same period the boreal lynx inhabited most of Scandinavia, Finland, the

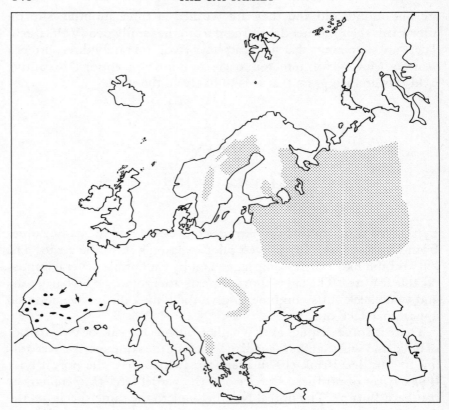

Baltic States and most of northern Asia with an arm stretching around
the Himalayas into Indo–China and another through central Europe to
the Pyrénées. Thus the two forms overlapped from the Massif Central
to the Pyrénées. The present distribution of the boreal lynx in Europe
includes parts of Norway and Sweden, Russia and parts of the Baltic
States and Finland, and an isolated population in the Carpathians. The
pardel lynx is found in a few parts of Spain and Portugal, the Carpathians and Greece and Albania.

   Lynxes disappeared from the British Isles in prehistoric times. They
were probably already becoming scarce but were pushed into extinction by Neolithic hunters. As the great European forests were felled
the lynxes retreated on all fronts. In Sweden, the lynx has been pushed
northwards so that it is now largely confined to an area above its
original northern limit, but lynx have been recently introduced to
parts of Yugoslavia, Germany, Switzerland and Italy (Smit and van
Wijngaarden, 1976).

Both lynxes live in areas of dense vegetation from the pine forests of Scandinavia to the marismas, maquis and oak woods of Spain. On the mountains of Spain, the lynxes depend on a good cover of under-growth, and where the oak and cork woods have been felled and replaced by conifer plantations there is no cover for the lynxes nor for the rabbits which are their principal prey (Heydt, 1971).

The life of the lynx is solitary in the main, each individual occupying its own range. A Swedish lynx has a central home range of about 300 square kilometres plus an outlying area which it visited at intervals and was shared with other lynxes. Occasionally adult lynxes will combine and hunt together. Haglund (1966) records two instances of lynxes apparently hunting in co-operation, one frightening a hare towards the other. This could have been a coincidental positioning but the adults shared the meal amicably. Lynx mark out their territories with sprays of urine, as does the domestic tom-cat, and by scratching tree trunks. Droppings are buried in the manner of domestic cats, except around the edges of the territory when they are left exposed as scent-marks.

The mating season starts at the end of February when males leave their normal ranges and wander in search of mates. Small packs may aggregate and fights break out between males, which may end in death. During this time, males advertise their presence with cater-wauling, high-pitched at first and ending in a moan. Gestation lasts for 63–73 days and the litter comprises up to four kittens. They are born in an underground lair, which may be an old badger sett, in a hollow tree, under dense undergrowth or even old white stork nests (Mountfort, 1958). The kittens are furred at birth but are blind and helpless. Development is slow. Eyes open at ten days and the first forays are made at five weeks but the kittens are suckled for at least nine months. Until they are weaned they retain their milk teeth and the claws do not develop. Thus kittens which lose their mothers during the first winter have little hope of surviving. The family does not normally split until the mother is ready to mate again. The natal coat has only the vaguest indications of spots which become definite at nine weeks. At fourteen weeks the adult hair and colouring are assumed (Kunc, 1970).

The winter habits of the northern lynx have been studied by track-ing animals over snow in the wilderness of Sweden (Haglund, 1966). Lynx were tracked for some 2,400 kilometres and their droppings and prey remains collected. They proved to be extremely mobile. An average day's travel was 19–20 kilometres along a wandering course which led the lynx through its range so that its hunting was never concentrated in one restricted place.

Lynxes are mainly nocturnal. They lie up during daylight hours and start hunting about an hour before nightfall. There is a further rest during the darkest part of the night but the night's activities are governed by the lynx's success in hunting. If hunting is successful it lies up more frequently.

The winter prey of the Swedish lynx is mainly deer, both reindeer and roe, hares and gamebirds, with foxes, rodents and small birds. Hares are the most important food overall and, in both the USSR and Canada, populations of lynxes fluctuate with those of hares. There is a swing to deer in late winter when hare and gamebird populations are at a minimum and deer are hampered by deep snow. Foxes are a surprising component of the lynx's diet. They are chased or dug out of their dens and tracks show that foxes try to keep clear of lynxes. Swedish hunters say that lynxes keep foxes away from a district and Spanish farmers have complained of increasing numbers of foxes after lynxes have disappeared.

The lynx finds its prey by sight and sound and is hampered by bad weather. It will lie in wait on a prominent hillock or bough and wait patiently for tell-tale signs of prey. Using hand-reared lynxes, Waldemar Lindeman (1955) showed that a rat can be detected at 75 metres, a rabbit at 300 metres and a roedeer at 600 metres. The lynx creeps towards a chosen prey showing all the patience, concentration and stealth of a domestic cat sneaking up on a sparrow. Alternatively it will wait for the prey to move towards it, perhaps by sitting alongside a hare trail. As with the domestic cat, the attack is explosive. The lynx throws itself forward in bounds, or leaps from its boulder or branch, hoping to catch its prey before it can react, so that the farther the lynx has to run, the less likely are its chances of success. Most kills are made from chases of less than 20 metres and at greater distances success drops drastically although deer are easier to catch in the snow than small mammals and birds because they break through the crust, but Haglund does record a hare caught after a chase of over 200 metres. Sixty-five to seventy per cent of deer attacked are caught, but only 35 per cent of hares and 25 per cent of gamebirds. Black grouse, capercaillie and hazel hens escape by vertical take-off but may be felled by the lynx leaping up and knocking them with its paws. Guy Mountfort (1958) tells how a Spaniard saw a lynx leap 2 metres in the air to catch a partridge.

Deer are brought down by the lynx leaping on their backs and aiming a bite at neck or throat which may kill through suffocation. Despite their large size and a rather slow death by suffocation or bleeding, no traces of struggle are left in the snow and Haglund concluded that the deer are sent into a state of shock by the attack. There is a similarity here with observations of lions and hunting dogs attacking large antelope in Africa and mention has already been made (p. 82) of David Livingstone's feelings when attacked by a lion.

The behaviour of the lynx with its kill is very much that of a predator of small animals in that it does not remain with the carcase of a large animal to take a second meal. It takes its fill, has a rest, then continues on its way. An animal the size of a hare is consumed entirely except for a few scraps but deer carcases are abandoned after a small part has been eaten. Flesh is usually from the thigh and back, according to Haglund but Valverde's (1957) study of the Spanish lynx led to a different conclusion. He found that the lynx is fastidious: eating only shoulder and neck of fawns, head and forepart of rabbits, and head, breast and wings of partridge. On rare occasions the lynx returns to the carcase for a further meal but it is more likely to seek new prey. This trait is a nuisance in Lapland where lynxes prey on the semi-domesticated reindeer. As the deer associate in herds the lynx can see several deer at once so it abandons the first kill, uneaten, and kills a second. The presence of an abundance of easily caught prey disrupts the predator's normal pattern of hunting in which hunger leads to searching for food and eating causes satiation. The result is a surplus-killing similar to that practised by foxes in unusual circumstances (p. 39).

According to Haglund, lynxes neither bury the remains of their prey, nor carry it from the scene of the attack to eat elsewhere, but contradictory accounts come from the Coto Doñana of the Guadal-quivir delta in Spain. Here, the usual pattern of behaviour is to carry the prey some distance—one lynx carried a young deer 150 metres, and the remains are buried, or partially covered in the case of large prey. A possible reason for the discrepancy in the food burying is that prey buried in the snow of a Swedish winter becomes frozen and lynxes are incapable of dealing with ice-hard meat.

Feral Dogs, Cats
and Ferrets

Feral animals are domesticated animals which have reverted, by accident or intent, to a wild existence. This category includes European populations of genets, mongooses, American mink and raccoons although none of these animals can be described as having been truly domesticated. Several carnivores have gone wild after being properly tamed and trained, in particular dogs, cats and ferrets. Where they have not wholly severed their links with man for food and shelter, they are better described as strays.

DOGS

Of the three, dogs are the least likely to go wild as they enjoy the closest ties with their owners. A feral dog is one stage farther removed from domestication than a stray (Nesbitt, 1975). It is a wild animal which shows fear of human beings. Feral dogs tend to be of larger breeds such as Alsatian, greyhound, Labrador and collie. When they go wild, their whole demeanour changes. The hindquarters are carried low so that the back is sloping and the gait becomes 'bouncy' when running, but the most characteristic feature of a feral dog is its furtive manner. Its head is held low and continually swings from side to side, both when running and when still (Burton, 1968). Sometimes it is difficult to recognize such an animal as a dog if no more than a fleeting glimpse is caught.

A feral dog is fully independent, although it may raid human habitation for scraps of other food. Little is known about the behaviour in the wild but there are records of dogs surviving for years by feeding on rabbits and hares. They lie up in nests of flattened grass which are changed every few days. Feral dogs are mainly solitary but they will breed in the wild and Nesbitt describes the behaviour of a pack of wild-born dogs in Illinois, USA.

CATS

The cat 'that walked by himself' is more likely to go feral than a dog. Cats hunt for themselves even when house pets, while a dog will rush after birds and rabbits ineffectually. The extent to which cats run wild is shown by the records from the Penrhyn Estates in North Wales. Between 1874 and 1902, ninety-eight polecats, thirteen pine martens and a massive 2,310 cats were shot on game preserves. Only a minority of these could be domestic cats out hunting as this was, and still is, a sparsely populated region. Nowadays there is a new source of feral cats. Animals bought on impulse as pets are often dumped in the country when their owners have become bored with them and many towns have large populations of feral cats which have strayed from their homes.

Nothing is known about the habits of feral cats in the country; they may be seen stalking through the fields and a nest of kittens may be found by accident. What is agreed by people who have met them is that feral cats are huge and ferocious. In size and weight, feral cats may be double the size of the average domestic cat, particularly if they become feral at an early age. It is also stated that by the second generation feral cats are as fierce when confronted as any wild cat and that in five generations they revert to the tabby coloration, but there is no hard evidence for these assertions.

FERRETS

The origin of the ferret is not known for certain. The first unequivocal record of ferrets seems to be that of Strabo who recorded in the first century B.C. that they were used to drive rabbits from their burrows in the Balearic Islands (Owen, 1969). Ferrets reached Germany and England in the thirteenth century, being introduced along with the rabbit. The uncertainty lies in whether ferrets are descended from European or steppe polecats. The bulk of evidence suggests the former, but Clifford Owen suggests that it was the idea of their use which spread and that the local form of polecat was domesticated wherever an animal was needed for bolting rabbits. Selective breeding has subsequently produced the ferret which is small (for easy access to burrows) and tame (for easy handling). Two strains have been developed: 'docile' for chasing but not killing rabbits and 'aggressive' for killing rats. The former are selected by destroying any young ferrets which have killer tendencies. Experimentally, ferrets have been shown to be less alert, less afraid of new situations or sudden changes and unafraid of man, when compared with European polecats (Poole, 1972).

Ferrets are paler than polecats and have larger patches of white over the eyes and around the ears. Albinos are common. They frequently stray, perhaps through failing to come out of a rabbit burrow before their owner's patience is exhausted. They can often be recaught because of their tameness and wild populations do not build up, perhaps through the activities of gamekeepers or from competition with polecats but they interbreed with wild polecats to produce fertile offspring. There are wild populations on Anglesey, the Isle of Man, some Scottish Islands and a few parts of mainland Britain and ferrets are widespread in New Zealand since their introduction in 1882 to combat rabbits.

The Fate of the Carnivores

The face of Europe has altered almost out of recognition in the last two millenniums. Once it was a wild continent of vast empty tracts of land with civilization only rubbing at the edges. There were unbroken stretches of forest, endless swamps and marshes and unconquered mountains. Now a human population, ever growing in numbers and technical ingenuity, has almost destroyed primaeval Europe. Instead there are unbroken stretches of sprawling human habitation and endless cultivated fields and pastures. The wildlife has changed as the environment has been altered. Some animals have flourished in the new man-made habitats; others succumbed or retreated to the last enclaves of wild Europe, pockets of land which are impossible or uneconomic to exploit.

The carnivores are among the animals hardest hit by the changed landscape. The expansion of human activity has hit them on several fronts. The destruction of forests and the spread of human settlement has altered much of the European carnivores' habitat, although some species have weathered the change. Forest-dwelling wild cats and pine martens have adapted to living on open moorland after the Scottish forests were felled, and foxes have become more abundant where agriculture has fragmented forests into separated woods and copses. By and large, however, agriculture and, of more importance, animal husbandry has brought carnivores into direct conflict with man. Their prey suffered from the spread of agriculture but, when they turned to livestock as an easy alternative, farmers were forced to retaliate. Later, the introduction of reliable firearms encouraged a new form of animal husbandry: the rearing of large numbers of game animals on preserves especially set aside for shooting. No carnivores were tolerated here, so gamekeepers joined stockmen and shepherds in the assault on the carnivores, now classed as vermin along with rats, mice and other competitors to man. At the same time, carnivores were being hunted

157

and trapped for their fur or merely for the pleasure of the chase. In furthering the latter activity, there has sometimes been a complete reversal in the normal treatment of carnivores, as when foxes have been deliberately introduced to an area to improve the hunting. Yet the overall trend has been a decline in the numbers of carnivores to a state where many are extinct in parts of their original range or rely on strict, active protection for their survival.

It is a sad fact that, in several instances, we are beginning to understand how carnivores live just as they are disappearing from our world. From some of the preceding accounts, it would seem as if carnivores are part of the country scene, particularly in the remoter areas of Europe. It is not surprising that the last pine marten in the London region was killed in 1883, but there are still bears, lynx, wolverines and wolves in the forests and mountains of Scandinavia. Haglund (1966) was able to study in detail the behaviour of wolverines and lynx in Sweden by following their tracks through the snow (p. 149). Yet he was studying rare animals. There were no more than one hundred wolverines and perhaps two or three times that number of lynx in the whole of Sweden. Wolves are even worse off. According to the IUCN Survival Service Commission there was, in November 1976, one wolf in Sweden—'the Padjelanta wolf'. In Norway the situation is not much better, with six wolves, and Finland can boast of only about fifty. Since then twenty have been killed. To discuss the role of a few dozen predators in terms of European ecology must be futile. The surviving animals may as well be in a zoo, at least they would be safer! Legal protection is not sufficient; the Lapps ignore Swedish protective legislation on the wolverine, for instance, and until very recently a bounty was still paid for the destruction of wolverines in Finland.

Perhaps it is surprising that any carnivores survive in Europe when one considers that almost every man's hand has been against them for several thousand years. Only two species have been removed from the European scene in historical times: the lion in classical times and the sable in the seventeenth century, but it is a wonder that there are any bears, wolves or lynxes left. Carnivores have often shown remarkable resilience in the face of persecution, but the greatest threat they face at present is habitat destruction rather than persecution. Protection may well have come too late for some European species as there are becoming too few places left where they can live. And if they can survive only in reserves with the assistance of man can they be considered really wild animals?

Before the natural habitats were destroyed, control of carnivores was a continuous and not always successful battle. As fast as an area was cleared of predators, they would spread in from neighbouring parts. The situation still occurs with common species, such as fox and stoat, and the Finnish wolf and brown bear populations are maintained by immigration from Russia. An added incentive for killing carnivores was the value of their pelts whose fine, dense underfur imparts a sensuous feel as well as warmth. Those most prized were marten, sable and ermine, while today lynx, otter, mink and Arctic fox are sought after, although few pelts come from wild European animals, except for the recent trend of using red fox pelts as trimmings. In mediaeval times, human social hierarchy extended to such matters as the furs appropriate to each rank. Ermine, sable and marten were available solely to royalty and the highest ranks of nobility, while the lower orders made use of lesser furs. In England, a decree of Edward III stated that yeomen and tradesmen could wear only fox, wild cat, lamb and rabbit.

Sport was, and is, another reason for killing carnivores. As well as the high sport of hunting as practised by the aristocracy with due pomp and ritual, there was the less organized hunting by commoners. 'Hunting the mart' was a popular pastime in Old England. The pine marten was chased by bands of men armed with sticks and stones and the polecat was hunted with dogs.

Fox hunting began centuries ago. By at least 350 B.C., the Persians were spearing foxes from horseback, as depicted on a seal of that era. In mediaeval England fox hunting was a minor recreation which filled a gap in winter when there was nothing better to hunt.

Fox hunting began to develop in earnest in the seventeenth century when increased clearance of land for agriculture favoured the fox and it became more abundant (Lloyd, H. G., 1976). Hunting was more organized in the eighteenth century and it reached its heyday in the first half of the nineteenth century. In England, fox hunting became a public pastime, either in participation or spectating, and it has been part of the social scene ever since. Hunts were also established in France but in Germany foxes were destroyed as vermin rather than preserved to be chased.

At present, fox hunting takes two basic forms. On one side the fox is an object to be chased and the pursuers are mounted. The fox need not be killed for a hunt to be successful. A good ride may satisfy some participants and the behaviour of the fox and the lie of its scent is important to others. In this form of hunting, foxes may be imported to bolster local stocks—the occasional wolf was imported accidentally (Lloyd, H. G., 1976)—and destruction of foxes for game preservation is frowned upon. In the other form of fox hunting, the aim is more directed at eliminating foxes. Such hunts take place in Scandinavia and the highland sheep districts of Britain. The hunters often work on foot as horses cannot cope with the rough terrain. It is sometimes said that the first form of hunting has been actually responsible for the preservation of foxes. This is true in some places where habitat supports a sparse fox population and trapping or shooting of foxes (discouraged by the hunt) could easily wipe them out. Elsewhere hunting, as with other forms of control, does no more than cream off the surplus. Lloyd (1976) gives an estimate of 50,000–100,000 foxes killed in Britain every year, and this has no long-term effect on the population.

The nineteenth century saw a further development in the war against carnivores. Accurate firearms allowed the shooting of game to flourish; gamebirds were reared artificially in large numbers, a great temptation for carnivores, so their keepers defended them with traps, snares and shot. Countless thousands of 'vermin' were massacred but, unless a species was already rare, there was every possibility that it could cope by breeding and immigration from areas where it was not persecuted.

To repeat what has been said before, it was the wholesale clearing of the land and destruction of habitat which has tipped the balance against the carnivores. In view of all these pressures on carnivore stocks, is there any ray of optimism which can dispel the gloomy prospects of extinction of the rarer species, summed up in the table opposite.

Polar bears have been suffering from man's recent exploitation of the Arctic. Men involved with military installations, mines and oil rigs have the opportunity to kill polar bears and they have been hunted from vehicles and aircraft, with the result that the polar bear became increasingly rare. In November 1973, the five nations with Arctic territory—Canada, Denmark (for Greenland), Norway (for Svalbard), USA and USSR—agreed to prohibit hunting. The USSR had already been protecting polar bears since 1956 and it has been reported that their numbers are increasing. Protection can be effective because there is still plenty of untouched habitat in the Arctic. The otter is more threatened than the polar bear, being classed by the IUCN as Endangered (in danger of extinction) rather than Rare (not endangered but at risk). Extinction is threatened in France, West Germany, Switzerland and Italy. Otters are legally protected in all but the last of these

THREATENED CARNIVORES OF EUROPE AND THEIR PROTECTION

	Status	Cause of Decline	USSR	Bulgaria	Romania	Yugoslavia	Hungary	Czechoslovakia	Poland	Greece	Italy	Austria	Portugal	Spain	France	Switzerland	Belgium	Netherlands	GDR	GFR	Denmark	Finland	Sweden	Norway	UK
Wolf	xxxxx	Persecution	H		PP	PP			PP				HR	HR								FP	FP	PP	
Brown bear	xxx	Habitat loss, persecution			H	H		FP	FP		PP	PP	HR	HR						HR		FP	HR	FP	
Polar bear	xxx	Persecution	HR						FP															HR	
European mink	xxxxx	Persecution, habitat loss	FP		H			FP	FP					FP	FP							(FP)	(FP)	HR	
Wolverine	xxxx	Persecution	H			H																PP	PP	H	
Otter	xxxxx	Habitat loss, persecution	HR		HR	PP		FP	FP		PP		PP	FP	FP	FP	PP	FP	FP	FP	FP	H	FP	PP	
Genet	xxx	Persecution							FP				PP		FP										
Wild cat	xxx	Persecution	FP	FP	H	HR		PP	PP		PP			FP	FP	FP	FP		FP	FP		FP	PP	HR	
Lynx	xxxx	Persecution, habitat loss	HR	HR		HR		PP	PP	HR	PP			FP	FP	FP	FP		FP	FP		FP	PP	FP	HR

Status: xxxxx—endangered; xxxx—vulnerable; xxx—rare. (Classification employed in IUCN Red Data Book).
Legislation: H=hunted freely; HR=hunting restrictions, e.g. quotas and closed season; PP=partial protection, e.g. in some regions; FP=full protection. (Based on Smit and van Wijngaarden, 1976.)

countries and in some other countries. (In Great Britain protection is extended in England and Wales but not in Scotland.) However, the main causes of decline among otters are water pollution, the clearing and damming of rivers, and disturbance. None of these factors are likely to change for the better in overpopulated, industrial Europe but the provision of otter havens, with strictly controlled public access and management, is allowing the otter to make a comeback in the Netherlands. As with the polar bear, the otter has enjoyed protection in the USSR since 1956 and, there being a large extant population and plenty of unspoilt habitat, otters are on the increase. Nevertheless, it is hard to be optimistic about the future of Europe's carnivores. Foxes have adapted to town life; weasels, even badgers, can thrive in the suburbs, but the majority stand little chance as human development continues to sprawl.

To add to the troubles facing the carnivores, several species are now being implicated in the spread of diseases that can infect man. It is ironic that two years after the Badger Act of 1973 had come into force and allowed the numbers of badgers to increase in parts of Great Britain (Neal, 1977), a programme of extermination was started in south-western England. In 1971 badgers were implicated in outbreaks of bovine tuberculosis in a limited area. Although a few foxes, rats and moles were also found to be infected, badgers are the most likely reservoir for infection passing to the cattle. Infection probably occurs from sputum, urine and faeces contaminating the pastures.

Badgers and tuberculosis is a local, and none too severe, problem. Far worse is the transmission of rabies to human beings. The classical carrier of rabies is the 'mad dog' but it is also transmitted by wolves, Arctic foxes and red foxes in Europe, and skunks and bats elsewhere. It is also found in several other European carnivores but the red fox is the main transmitter of the disease to man.

Rabies is a particularly nasty disease because its effects are excruciating and, once the disease has developed, it is fatal. Some parts of Europe had been clear of rabies. The disease was eliminated from Britain in 1903 and is currently absent from wild animals in Denmark, Norway, Sweden, Finland, Italy, Spain and Portugal but, in 1939 or 1940, rabies was observed in Polish foxes and, in the succeeding years, the disease has spread westwards on a broad front at a rate of 20–60 kilometres a year (Lloyd, H. G., 1976). Intensive slaughter of foxes has failed to stem the current spread through France; indeed, it may be hastened because the social life of the survivors is disrupted and they disperse over larger distances. Denmark was able to throw the disease back by establishing a *cordon sanitaire* across the narrow neck of the country, and insular Britain is keeping the disease out by strict quarantine of imported animals. Rabies is such a terrible disease that no efforts are too great to prevent its occurrence but it must be a matter of regret that so many badgers and other mammals are killed along with the foxes in the control campaigns.

Bibliography

Anon. (1971). Classified notes of Norfolk Mammal Report. *Trans. Norfolk Naturalists Trust* **22**, 362–76.

——(1974). Hermann Göring's secret weapon. *Wildlife* **16**, 375.

Apfelbach, R. (1973). Olfactory sign stimulus for prey selection in polecats (*Putorius putorius* L.) *Z. Tierspsychol* **33**, 270–3.

Balliet, R. F. and Schusterman, R. J. (1971). Underwater and aerial visual acuity in the Asian 'clawless' otter (*Amblonyx cinerea cineria*). *Nature* **234**, 305–6

Bannikov, A. G. (1964). Biologie du chien viverrin en URSS. *Mammalia* **28**, 1–39.

—— (1967). Bears and burunduks of Barguzinski. *Animals* **9**, 656–60.

Baranovskaya, T. N. and Kolosov, A. M. (1935). Food habits of the fox (*Vulpes vulpes* L.). *Zool. Zhurnal* **14**, 523–50.

Barlow, C. T. (1949). *The Countryman* **39** (2), 282.

Batten, M. (1920). *Habits and Characteristics of British Wild Animals.* Chambers.

Bengston, S. (1972). Reproduction and fluctuations in the size of duck populations at Lake Myvatn, Iceland. *Oikos* **23**, 35–58.

Bourne, W. R. P. (1978). Mink and wildlife. *BTO News* **91**, 1–2.

Braestrup, F. W. (1941). A study on the Arctic fox in Greenland *Medd. Gron.* **131**, 1–101.

Brink, F.-H. van den (1967). *A Field Guide to the Mammals of Britain and Europe.* Collins.

——(1970). Distribution and specification of some Carnivores. *Mammal Rev.* **1**, 67–78.

Brown, J. H. and Lasiewski, R. C. (1972). The metabolism of weasels: the cost of being long and thin. *Ecology* **53**, 939–43.

Burghardt, G. M. and Burghardt, L. S. (1972). Notes on the behavioural development of two female black bear cubs: the first eight months. IUCN Publ. new series **23**, 207–20.

Burrows, R. (1968). *Wild Fox.* David and Charles.

Burton, M. (1955). Introducing Jennie. *Illustrated London News 1955*, 702.

——(1968). *Wild Animals of the British Isles.* Warne.

Chanin, P. R. F. (1976). The ecology of the feral mink (*Mustela vison* Schreber) in Devon. Unpublished thesis, Exeter University.

Chirkova, A. F. (1953). Dynamics of fox numbers in Voronezh Province and forecasting fox harvests. *Trans. Russian Game Reports* **3**, 50–69.

Churcher, C. S. (1959). The specific status of the New World red fox. *J. Mammal.* **40**, 513–20.

Cocks, A. H. (1881). Notes on the breeding of the otter. *Proc. Zool. Soc. Lond.* **1881**, 249–50.

Condé, B. *et al.* (1972). Le régime alimentaire du chat forestier (*F. silvestris* Schr.) en France. *Mammalia* **36**, 112–19.

Corbet, G. (1966). *The Terrestrial Mammals of Western Europe*. Foulis.

Corbet, G. B. and Southern, H. N. (ed.) (1977). *The Handbook of British Mammals*. Blackwell.

Couturier, M. A. J. (1954). *L'ours brun*. Grenoble.

Cowan, I. McT. (1972). The status and conservation of bears (Ursidae)- of the world. IUCN Publ. new series **23,** 343–67.

Cozens, J. (1959). *The Countryman* **56** (2), 293–4.

Curry-Lindahl, K. (1972). The brown bear (*Ursus arctos*) in Europe: decline, present distribution, biology and ecology. IUCN Publ. new series 23.

Cuthbert, J. H. (1973). The origin and distribution of feral mink in Scotland. *Mammal Rev.* **3**, 97–103.

Darling F. F. and Boyd, J. M. (1964). *The Highlands and Islands*. Collins.

Davies, G. (1936). Distribution of the badger (*Meles meles*) around Denbigh, with notes on its food and habits. *J. Anim. Ecol.* **5**, 97–104.

Davis, D. D. (1945). The shoulder architecture of bears and other carnivores. *Fieldiana Zool.* **31**, 285–305.

Day, M. G. (1968). Food habits of British stoats (*Mustela erminea*) and weasels (*Mustela nivalis*) *J. Zool.* **155**, 485–97.

Dean, F. (1949). *The Countryman* **39**,(2), 281–2.

Deansley, E. (1943). Delayed implantation in the stoat (*Mustela nivalis*). *Nature* **151**, 365–6.

——(1944). The reproductive cycle of the female weasel (*Mustela nivalis*). *Proc. Zool. Soc. Lond.* **114**, 339–49.

Drabble, P. (1969). *Badgers at My Window*. Pelham Books.

Dunn, E. (1977). Predation by weasels (*Mustela nivalis*) on breeding tits (*Parus* spp.) in relation to the density of tits and rodents. *J. Anim. Ecol.* **46**, 633–52.

Dunwell, R. and Killingray, A. (1969). The distribution of badger setts in relation to the geology of the Chilterns. *J. Zool.* **158,** 204–8.

East, K. and Lockie, J. M. (1964). Observations on a family of weasels (*Mustela nivalis*) bred in captivity. *Proc. Zool. Soc. Lond.* **143**, 359–63.

Ellerman, J. R. and Morrison-Scott, T. C. S. (1951). The Checklist of Palaearctic and Indian Animals. British Museum (Natural History).

Elmhirst, R. (1938). Food of the otter in the marine littoral zone. *Scott. Nat.* **1938**, 99–102.

Englund, J. (1965). Some aspects of reproduction and mortality rates in Swedish foxes (*Vulpes vulpes*), in 1961–3 and 1966–9. *Viltrevy* **8**, 1–82.

Erlinge, S. (1967). Food habits of the fish-otter, *Lutra lutra* L., in south Swedish habitats. *Viltrevy* **4**, 372–433.

——(1968a). Territoriality of the otter. *Oikos* **19**, 81–98.

——(1968b). Food studies of captive otters. *Oikos* **19**, 259–70.

——(1969). Food habits of the otter *Lutra lutra* L. and mink *Mustela vison* Schreber in a trout water in S. Sweden. *Oikos* **20**, 1–7.

——(1972). Interspecific relations between otter *Lutra lutra* and mink *Mustela vison* in Sweden. *Oikos* **23**, 327–35.

Ewer, R. F. (1968). *Ethology of Mammals*. Logos.

——(1973). *The Carnivores.* Weidenfeld and Nicolson.

Fairley, J. S. (1969).. The fox as a pest of agriculture. *Irish Nat. J.* **16**, 216–19.

Fog, M. (1969). Studies on the weasel (*Mustela nivalis*) and the stoat (*Mustela erminea*) in Denmark. *Dan. Rev. Game Biol.* **6**, 1–14.

Fox, M. W. (1971a). Socio-infantile and socio-sexual signals in Canids: a comparative and ontogenetic study. *Z. Tierpsychol.* **28**, 185–210.

——(1971b). Possible examples of high order behaviour in wolves. *J. Mammal.* **52**, 640–1.

Gangloff, B. and Ropartz, P. (1972). Le repertoire comportmental de la genette *Genetta genetta* (Linné). *Terre et vie* **26**, 489–560.

Gerell, R. (1967). Food selection in relation to habitat in mink (*Mustela vison* Schreber) in Sweden. *Oikos* **18**, 233–46.

——(1969). Activity patterns of the mink (*Mustela vison* Schreber) in southern Sweden. *Oikos* **20**, 451–60.

——(1970). Home ranges and movements of the mink (*Mustela vison* Schreber) in southern Sweden. *Oikos* **21**, 160–73.

——(1971). Population studies on the mink (*Mustela vison* Schreber) in southern Sweden. *Viltrevy* **8**, 83–114.

Goethe, F. (1940). Beiträge zur Biologie des Iltis. *Z. Saugetierk.* **15**, 180–223.

Haglund, B. (1966). De Stora Rovdjurens vintervanor 1. (English summary) *Viltrevy* **4**, 81–308.

Hamilton, E. (1896). *The Wild Cat of Europe (Felis catus)*. R. H. Porter.

Harrington, C. R. (1965). The life and status of the polar bear. *Oryx* **8**, 169–76.

——(1968). Denning habits of the polar bear (*Ursus maritimus* Phipps). *Canad. Wildl. Service Rept.* Ser. No. 5, 1–30.

Harris, C. J. (1968). *Otters. A study of the Recent Lutrinae*. Weidenfeld and Nicolson.

Harris, S. (1977). Distribution, habitat utilization and age structure of a suburban fox (*Vulpes vulpes*) population. *Mammal Rev.* **7**, 25–39.

Hartman, L. (1964). The behaviour and breeding of captive weasels (*Mustela nivalis* L.). *N. Z. Jl. Sci.* **7**, 147–56.

Henshaw, R. E. *et al.* (1971). Peripheral thermoregulation: foot temperature in two Arctic carnivores. *Science* **175**, 988–90.

Herbert, W. (1969). *Across the Top of the World*. Longmans.

Hewson, R. (1969). Couch building by otters *Lutra lutra*. *J. Zool.* **159**, 524–7.

——(1973). Food and feeding habits of otters *Lutra lutra* at Loch Park, north-east Scotland. *J. Zool.* **170**, 159–62.

——and Healing, T. D. (1971). The stoat *Mustela erminea* and its prey. *J. Zool.* **164**, 239–46.

Heydt, J. G. (1971). Quoted in *World of Wildlife*, Orbis, **5**, 12–16.

Hinton, H. E. and Dunn, A. M. S. (1967). *Mongooses: their natural history and behaviour*. Oliver and Boyd.

Hurrell, H. G. (1963). *Pine martens*. Sunday Times.

Jefferies, D. J. (1975). Different activity patterns of male and female badgers (*Meles meles*) as shown by road mortality. *J. Zool.* **177**, 504–6.

Jewell, P. A. (1966). The concept of home range in mammals. *Symp. Zool. Soc. Lond.* **18**, 85–110.

Jonkel, C. J. *et al.* (1972). Further notes on polar bear denning habits IUCN Publ. new series **23**, 172–80.

Kaikusalo, A. (1971). On the breeding of the arctic fox. (English summary.) *Suomen Riista* **23**, 7–16.

King, C. M. (1975a). The sex ratio of trapped weasels (*Mustela nivalis*). *Mammal Rev.* **5**, 1–8.

——(1975b). The home range of the weasel (*Mustela nivalis*) in an English woodland. *J. Anim. Ecol.* **44**, 639–68.

Kistchinski, A. A. and Uspenski, S. M. (1972). Immobilization and tagging of polar bears in maternity dens. IUCN Publ. new series **23**, 172–80.

Kleiman, D. (1966). Scent marking in the Canidae. *Symp. Zool. Soc. Lond.* **18**, 167–77.

Kleinenberg, S. E. *et al.* (1964). *Beluga (Delphinapterus leucas): investigation of the species*. Israel Program for Scientific Translation.

Knott, P. (1960). *Der Vielfrass*. A. Ziemsin Verlag.

Krebs, J. R. (1970). Regulation of numbers in the great tit (Aves: Passeriformes). *J. Zool.* **162**, 317–33.

Kruuk, H. (1964). Predators and antipredator behaviour of the black-headed gull (*Larus ridibundus* L.). *Behaviour*, Suppl. xi, 1–129.

——(1978). Spatial organisation and territorial behaviour of the European badger *Meles meles*. *J. Zool.* **184**, 1–19.

Kunc, L. (1970). Breeding and rearing the northern lynx *Felis l. lynx* at Ostrava zoo. *Internat. Zool. Yrbk.* **10**, 38–40.

Langley, P. J. W. and Yalden, D. W. (1977). The decline of the rarer carnivores in Great Britain during the nineteenth century. *Mammal Rev.* **7**, 95–116.

Layne, J. N. and McKeon, W. H. (1956). Some aspects of red fox and gray fox reproduction in New York. *New York Fish and Game Journal* **3**, 44–74.

Leeuw, A. de (1957). Die wildkatze. *Merkol. Niederwild. dt. Jagdschutzverb.* 16.

Lentfer, J. W. (1972). Polar bear—sea ice relationships. IUCN Publ. new series **23**, 165–71.

Lever, R. J. A. W. (1959). The diet of the fox since myxomatosis. *J. Anim. Ecol.* **28**, 359–75.

Leyhausen, P. (1964). Communal organisation of solitary animals. *Symp. Zool. Soc. Lond.* **14**, 249–63.

Liers, E. E. (1951). Notes on the river otter (*Lutra canadensis*). *J. Mammal.* **32**, 1–9.

Lindeman, W. (1955). Über die jugendentwicklung beim luchs (*Lynx l. lynx* Kerr) und bei der wildkatze (*Felis S. silvestris* Schreb.). *Behaviour* **8**, 1–45.

Lingard, J. (1965). Rearing a family of weasels. *Animals* **5**, 495–500.

Linn, I. (1962). *Weasels*. Sunday Times.

Lloyd, H. G. (1968). The control of foxes. *Ann. Appl. Biol.* **61**, 334–45.

——(1975). The red fox in Great Britain. In *The Wild Canids* (ed. M. W. Fox). Van Nostrand Reinhold.

——(1976). Wildlife rabies in Europe and the British situation. *Trans. Roy. Soc. of Trop. Med. and Hygiene* **70**, 179–87.

Lloyd, J. R. (1968). Factors affecting the emergence time of the badger (*Meles meles*) in Britain. *J. Zool.* **155**, 223–7.

Lockie, J. D. (1961). The food of the pine marten *Martes martes* in West Ross-shire, Scotland. *Proc. Zool. Soc. Lond.* **136**, 187–95.

——(1964). Distribution and fluctuations of the pine marten *Martes martes* L. in Scotland. *J. Anim. Ecol.* **33**, 349–56.

——(1966). Territory in small carnivores. *Symp. Zool. Soc. Lond.* **18**, 143–65.

Lund, H. M-K. (1962). Feeding of the fox in Norway. *Medd. St. Viltundersok* **12**, 1–75.

Lyall-Watson, M. (1963). A critical re-examination of food 'washing' behaviour in the raccoon (*Procyon lotor* Linn.). *Proc. Zool. Soc. Lond.* **141**, 37–193.

Macdonald, D. W. (1976). Food catching by red foxes and some other carnivores. *Z. Tierpsychol.* **42**, 170–185.

——(1977). On food preference in the red fox. *Mammal Rev.* **7**, 7–23.

Macintyre, D. (1936). *Wildlife of the Highlands*. Philip Allen.

——(1950). Habits of the otter. *Field* **196**, 546.

Mackenzie, O. H. (1924). *A Hundred Years in the Highlands*. (Popular Edition.) Edward Arnold.

Malone, E. V. (1965). *The Countryman* **65** (1), 100.

Matthews, L. H. (1952). *British Mammals*. Collins.

Mead, C. J. and Pepler, G. R. M. (1975). Birds and other animals at sand martin colonies. *Brit. Birds* **68**, 88–9.

Mech, L. D. (1970). *The Wolf*. Amer. Mus. Nat. Hist.

Meyer-Holzapfel, M. (1968). Breeding of the European wildcat *Felis S. sylvestris* at Berne Zoo. *Internat. Zool. Yrbk.* **8**, 31–8.

Mikkola, H. (1974). The raccoon dog spreads to western Europe. *Wildlife* **16**, 344–5.

Millais, J. G. (1904). *Mammals of Great Britain and Ireland*. Longmans Green.

Miller, G. S. (1912). *Catalogue of the Mammals of Western Europe*. British Museum (Nat. Hist.).

Moors, P. J. (1975). The food of weasels (*Mustela nivalis*) on farmland in north-east Scotland. *J. Zool.* **177**, 455–61.

Mountford, G. (1958). *Portrait of a Wilderness*. Hutchinson.

Mysterud, I. (1973). Behaviour of the brown bear at moose kills. *Norwegian J. Zool.* **21**, 267–72.

Nansen, F. (1898). *Farthest North*. Newnes.

Neal, E. (1948). *The Badger*. Collins.

——(1962). *The Countryman* **63** (2), 432.

——(1972). The national badger survey. *Mammal Rev.* **2**, 55–64.

——(1977). *Badgers*. Blandford.

——and Harrison, R. J. (1958). Reproduction in the European badger (*Meles meles* L.) *Trans. Zool. Soc. Lond.* **29**, 67–120.

Nesbitt, W. H. (1975). Ecology of a feral dog pack on a wildlife refuge. In *The Wild Canids*. Ed. M. W. Fox. Van Nostrand Reinhold, pp. 391–6.

Nethersole-Thompson, D. and Watson, A. (1974). *The Cairngorms*. Collins.

Notini, G. (1948). Quoted in *The Handbook of British Mammals*. Ed. H. N. Southern. Blackwell, 1964.

Novikov, G. A. (1939). *The European mink*. (English summary.) Leningrad.

——(1956). *Carnivorous mammals of the fauna of the USSR*. Israel Program for Scientific Translation, 1962.

Nyholm, E. S. (1959). Stoats and weasels in their winter habitat. *Suomen Riista* **13**, 106–16. Translation in *Biology of Mustelids*. British Library Lending Division, 1975, pp. 118–31.

——(1971). Ecological observations on the snow hare (*Lepus timidus* L.) on the islands of Krunnit and Kuusamo. *Suomen Riista* **23**, 115.

Ognev, S. I. (1931). *Mammals of Eastern Europe and Northern Asia*. Vol. 2 *Carnivora, Fissipedia*. Israel Program for Scientific Translation, 1962.

Ondrias, J. C. (1961). Comparative osteological investigation on the front limbs of European Mustelidae. *Ark. Zool.* **13**, 311–20.

Osterholm, H. (1964). The significance of distance receptors in the feeding behaviour of the fox, *Vulpes vulpes* L. *Acta Zool. Fenn.* **106**, 1–31.

Owen, C. (1965). Pine marten, polecat, stoat, weasel. *Animals* **7**, 128–33.

——(1969). The domestication of the ferret. In *The Domestication and Exploitation of Plants and Animals*. Ed. P. J. Ucko and G. W. Dimbleby. Duckworth.

Paget, R. J. and Middleton, A. L. V. (1974). Some observations on the sexual activities of badgers (*Meles meles*) in Yorkshire in the months December to April. *J. Zool.* **173**, 256–60.

Parovshchikov, V. Ya. (1963). A contribution to the ecology of *Mustela nivalis* Linnaeus, 1766, of the Archangel'sk north. Translation in *Biology of Mustelids*. British Library Lending Division, 1975, pp. 79–83.

Pedersen, A. (1966). *Polar Animals*. Harrap.

Pentecost, E. L. C. (1963). The plight of a weasel. *Field* 25 July, 167.

Perrins, C. M. (1965). Population fluctuations and clutch size in Great Tits, *Parus major*. *J. Anim. Ecol.* **34**, 601–47.

Perry, R. (1973). *Polar worlds*. David and Charles.

Pocock, R. I. (1911). Some probable and possible instances of warning characteristics among insectivorous and carnivorous mammals. *Ann. Mag. Nat. Hist.* **8**, 750–7.

Poole, T. B. (1966). Aggressive play in polecats. *Symp. Zool. Soc. Lond.* **18**, 23–44.

——(1972). Some behavioural differences between the European polecat, *Mustela putorius*, the ferret, *M. furo*, and their hybrids. *J. Zool.* **166**, 25–35.

Pulliainen, E. (1965). Studies on the wolf (*Canis lupus* L.) in Finland. *Ann. Zool. Fenn.* **2**, 215–59.

Reichstein, H. (1957). Schädelvariabilität europäischer Mausewiesel (*Mustela nivalis* L.) und Hermeline (*Mustela erminea* L.) in Beziehung zu Verbreitung und Geschlecht. *Z. Säugetierk.* **22**, 151–82.

Rothschild, M. (1957). Note on change of pelage in the stoat. *Proc. Zool. Soc. Lond.* **128**, 602.

Rust, C. C. (1962). Temperature as a modifying factor in the spring pelage change of short-tailed weasels. *J. Mammal.* **47**, 602–12.

Ryszkowski, L. *et al.* (1971). Operation of predators in a forest and cultivated fields. *Ann. Zool. Fenn.* **8**, 160–8.

——(1973). Trophic relationships of the common vole in cultivated fields. *Acta. Theriol.* **18**, 125–65.

Schmook, A. (1960). *Der Fuchs*. Otto Verlag Thun.

Scholander, P. F. *et al.* (1950). Body insulation of some Arctic and tropical mammals and birds. *Biol. Bull.* **99**, 225–36.

Sinclair, W. *et al.* (1974). Aerial and underwater visual activity in the mink *Mustela vison* Schreber. *Anim. Behav.* **22**, 965–74.

Skoog, P. (1970). The food of the Swedish badger. *Viltrevy* **7**, 1–120.

Smit, C. J. and Wijngaarden, A. van (1976). Threatened mammals in Europe. Nature and Environment Series No. 10. Council of Europe.

Sokolov, A. S. and Sokolov, I. I. (1970). Some specific features of the locomotor organs of river- and sea-otters. (English summary.) *Byull. Mosk. Obshch. Ispȳt. Prir. Otd. Biol.* **75**, 5–17.

Sokolov, I. I. and Sokolov, A. S. (1971). Some features of the locomotor organs of *Martes martes* L. associated with its mode of life. *Byull. Mosk. Obshch. Ispȳt. Prir. Otd. Biol.* **76**, 40–51.

——(1972). Some specific features of the locomotor organs of the badger as a burrower. (English summary.) *Byull. Mosk. Obshch. Ispȳt. Prir. Otd. Biol.* **77**, 19–28.

Southern, H. N. and Watson, J. S. (1941). Summer food of the red fox (*Vulpes vulpes*) in Great Britain: a preliminary report. *J. Anim. Ecol.* **10**, 1–11.

Stephens, M. N. (1957). *The Otter Report*. UFAW.

Stuewer, F. W. (1943). Raccoons: their habits and management in Michigan. *Ecol. Monogr.* **13**, 203–57.

Stonorov, D. and Stokes, A. W. (1972). Social behaviour of the Alaska brown bear. IUCN Publ. new series **23**, 232–42.

Tarasoff, F. J. *et al.* (1972). Locomotory patterns and external morphology of the river otter, sea otter and harp seal (Mammalia). *Canad. J. Zool.* **50**, 915–29.

Tembrock, G. (1957). Das Verhalten des Rotfuchses. *Handl. zool. Berlin* **8**, 1–20.

Thesiger, W. (1959). *Arabian Sands*. Longmans Green.

Tschanz, B. *et al.* (1970). Das Informationssystem bei Braunbären. *Z. Tierpsychol.* **27**, 47–72.

Valverde, J. A. (1957). Notes ecologiques sur le lynx d'Espagne *Felis lynx pardina* Temminck. *Terre et vie* **104**, 51–67.

Vereschagin, N. K. (1959). *The Mammals of the Caucasus*. Israel Program for Scientific Translation, 1967.

Vesey-Fitzgerald, B. (1942). *A Country Chronicle*. Chapman and Hall.

Vincent, R. E. (1958). Observations of red fox behaviour. *Ecology* **39**, 755–7.

Volf, J. (1968). Breeding of a wild cat *Felis s. sylvaticus* at Prague Zoo. *Internat. Zool. Yrbk.* **8**, 38–42.

Walker, D. R. G. (1972). Observations on a collection of weasels (*Mustela nivalis*) from estates in south-west Hertfordshire. *J. Zool.* **166**, 474–80.

Walker, E. P. (1964). *Mammals of the World*. Vol. ii. Johns Hopkins Press.

Walton, K. C. (1970). The polecat in Wales. *Welsh Wildlife in Trust*. West Wales Naturalists Trust, 98–108.

Watson, A. (1961). Wildcat making a bed in the open. *Scott. Nat.* **70**, 85–6.

Weir, V. and Banister, K. E. (1971). The food of the otter in the Blakeney area. *Trans. Norf. Norw. Nat. Hist. Soc.* **22**, 377–82.

White, T. H. (1954). *The Book of Beasts*. Jonathan Cape.

Wijngaarden, A. van and Peppel, J. van de (1964). The badger (*Meles meles*) in the Netherlands. *Lutra* **6**, 1–60.

Wüsterhube, C. (1960). Beiträge zur kenntnis besonders Spiel und Benterverhaltens einheimischer Mustelidae. *Z. fur Tierpsychol.* **17**, 579–613.

Yurgenson, P. B. (1947). Sexual dimorphism in feeding as an ecological adaptation of a species. Translation in *Biology of Mustelids*. British Library Lending Division, 1975, pp. 79–83.

Zeuner, F. E. (1963). *A History of Domesticated Animals*. Hutchinson.

Zunino, F. and Herrero, S. (1972). The status of the brown bear (*Ursus arctos*) in Abruzzi National Park, Italy. 1971. *Biol. Conserv.* **4**, 263–72.

Index

Aardwolf 11

American mink, breeding 95; colour 94; compared with European mink 92, with otter 96, with stoat 95; delayed implantation 95; den 95; distribution 94; food 95–6; gestation 95; habitat 94–5; hunting 96; induced ovulation 95; litter 95; measurements 94; scent 95; spraint 95; territory 95; transients 95

Aquatic adaptations, European mink 92–3; otter 125–6, 128–9, 131–2; polar bear 68

Arctic fox, breeding 48; cacheing 47–8; colour 44; den 46; dispersal 46; distribution 45; food 46–7; gestation 48; habitat 45; hunting 46–7; litter 48; measurements 45; relations with polar bear 47; scent 48; surplus killing 47; voice 48; weaning 48

Ashby, Eric 41

Asian polecat, *see* Steppe polecat

Baculum 16

Badger, bedding 19, 118; bite 15; breeding 122–4; colour 113; delayed implantation 123; den, *see* Sett; distribution 116; dung pits 118, 124; emergence 119; food 120–2; funeral 18, 125; gestation 123; habitat 116–17; hearing 114; home range 124; induced ovulation 122; jaw 15, 115; limbs 114; litter 123; measurements 113; persecution and protection 113, 116, 162; play 122–3; relations with fox 32, with wolf 56; scent 114–15, 122, 124–5; skeleton 114; skull 114; smell 114, 121; teeth 12–13, 116, 120; territory 125; tuberculosis 116, 162; vision 114; voice 115; weaning 123; winter sleep 119

Bears, *see* Brown bear and Polar bear; feet 16; limbs 14–15; skull 14; teeth 12–13

Beech marten, breeding 109; colour 108; delayed implantation 109; den 109; distribution 108–9; food 109; gestation 109; habitat 108–9; litter 109; measurements 108

Bingley, William 144, 145

Bobcat 147

Breeding 18, 22–3; American mink 95; Arctic fox 48; badger 122–4; beech marten 109; brown bear 64–5; Canidae 29; Egyptian mongoose 142; European mink 94; European polecat 100; feline genet 140; golden jackal 58; lynx 149; marbled polecat 103; otter 135–6; pine marten 107–8; polar bear 70; raccoon 74–5; raccoon dog 51; red fox 22, 38, 39–42, 44; steppe polecat 101–2; stoat 90–1; weasel 84–5; wild cat 146; wolf 54–5; wolverine 112

Brown bear, breeding 64–5; cacheing 63; colour 60; danger to man 63; delayed implantation 64; den 64; distribution 60–1; facial expression 63; food 62–3; habitat 60–1; home range 62; limbs 61–2; litter 64–5; measurements 60; persecution and protection 60, 161; scent 62; teeth 62; weaning 65; winter sleep 64–5

Burrow, *see* Den

Burrows, Roger 44

Burton, Jane 41

Cafer cat 144, 145

Cacheing, Arctic fox 47–8; brown bear 63; European mink 94; lynx 151; red fox 37–8; weasel 82; wolverine 112

Camville, Gerard 159

Canidae 11, 29; feet 16; teeth 12–13

Canoidea 11

Caracal 147

Carnassials 12, 14

Carnivores, European 11, 26–7; limbs 16

Charming, red fox 36–7; stoat 89–90

Classification 11, 26

Colour, American mink 94; Arctic fox 44; badger 113; beech marten 108; brown bear 60; Egyptian mongoose 141; European mink 92; European polecat 97–8; feline genet 138; ferret 155; golden jackal 56; lynx 147; marbled polecat 102; otter 126–7; pine marten 104; polar bear 66; raccoon 74; raccoon dog 48; red fox 30; steppe polecat 101; stoat 85–8; weasel 78; wild cat 144; wolf 52; wolverine 110

Couch 133

Danger to man, brown bear 63; polar bear 70–1; wild cat 144; wolf 53

Delayed implantation 23–4; American mink 95; badger 123; beech marten 109; brown bear 64; otter 136; pine marten 107; polar bear 70; stoat 90–1; Ursidae 59; wolverine 112

Den, American mink 95; Arctic fox 46; badger, *see* Sett; beech marten 109; brown bear 64; European polecat 100; feline genet 140; golden jackal 58; lynx 149; otter 133; pine marten 107; polar bear 69–70; raccoon 74; raccoon dog 50; red fox 32; steppe polecat 101; weasel 83; wild cat 146; wolf 54; wolverine 112

Dew claw 16; Canidae 29; Felidae 143

Dispersal, Arctic fox 46; pine marten 105; red fox 22, 42

Distribution, American mink 94; Arctic fox 45; badger 116; beech marten 108–9; brown bear 60; Egyptian mongoose 140–1; European mink 93; European polecat 98; feline genet 138; golden jackal 56–7; lynx 147–9; marbled polecat 102; Mustelidae 77; otter 127; pine marten 104; polar bear 65–7; Procyonidae 73; raccoon 74; raccoon dog 48–9; red fox 30–1; steppe polecat 101; stoat 86; Ursidae 59; Viverridae 137; weasel 79; wild cat 144–5; wolf 52–3; wolverine 110–1

Dung pits 118, 124

Earth, *see* Den

Egyptian mongoose, breeding 142, colour 141; distribution 140–1; food 142; habitat 142; litter 142; measurements 141; scent 142; weaning 142

Ermine 86–7, 159

Europe, carnivore fauna 11, 26–7; defined 25

European mink, aquatic adaptation 92–3; breeding 94; cacheing 94; colour 92; compared with American mink 92; distribution 93; food 93–4; gestation

174